# THE CONSTRUCTION OF BUILDINGS

R. BARRY
A.R.I.B.A.

*M... ...tions,*
*Fi... ...ishes*

First published, 1960
Reprinted with SfB headings, 1963
Reprinted, 1965
Reprinted, 1967
Reprinted, 1969
Second edition (metric), 1970

ISBN 0 258 96798·6

REPRODUCED PHOTOLITHO IN GREAT BRITAIN BY
J. W. ARROWSMITH LTD. WINTERSTOKE ROAD, BRISTOL

*CONTENTS*


| | | |
|---|---|---|
| ONE | MASONRY .. .. .. .. .. .. | 1 |
| TWO | WINDOWS .. .. .. .. .. .. | 24 |
| THREE | WINDOW SILLS .. .. .. .. .. | 52 |
| FOUR | DOORS .. .. .. .. .. .. | 60 |
| FIVE | FIREPLACES – FLUES – HEARTHS .. .. .. | 85 |
| SIX | PARTITIONS .. .. .. .. .. | 106 |
| SEVEN | STAIRS .. .. .. .. .. .. | 116 |
| EIGHT | INTERNAL FINISHES AND EXTERNAL RENDERING | 128 |
| | INDEX .. .. .. .. .. .. | 138 |

## ACKNOWLEDGMENTS

The author has had advice and assistance from many of his colleagues at the School of Building, Brixton, in the preparation of this book, and in particular from Mr. A. Cleveland, B.Sc., who read the manuscript and made helpful suggestions.

## FIRST METRIC EDITION: NOTE

For linear measure all measurements are shown in either metres or millimetres. A decimal point is used to distinguish metres and millimetres, the figures to the left of the decimal point being metres and those to the right millimetres. To save needless repetition, the abbreviations 'm' and 'mm' are not used, with one exception. The exception to this system is where there are at present only metric equivalents such as those published with the Building Regulations, in decimal fractions of a millimetre. Here the decimal point is used to distinguish millimetres from fractions of a millimetre, the figures to the left of the decimal point being millimetres and those to the right being fractions of a millimetre. In such cases the abbreviation 'mm' will follow the figures (e.g. 152·4mm).

April 1970 R. BARRY

CHAPTER ONE

# MASONRY

## SfB (21) Walls: External, load bearing: Masonry

Before the industrial revolution the majority of permanent buildings in hill and mountain districts, and most large buildings in lowland areas in this country were built of natural stone. At that time the supply of stone from local quarries was adequate for the buildings of the small population of this country. The increase in population that followed the industrial revolution was so great that the supply of sound stone was quite inadequate for the new buildings being put up. Coal was cheap, the railway spread throughout the country and cheap mass-produced bricks largely replaced stone as the principal material for the walls of all but the larger buildings.

Because natural stone is expensive it is principally used today as a facing material bonded or fixed to a backing of brickwork or concrete. Many of the larger civic and commercial buildings are faced with natural stone because of its durability, texture, colour and sense of permanence.

Natural stone is also used as the outer skin of cavity walls for houses in areas where local quarries can supply stone at reasonable cost.

Because natural stone is an expensive material and in quarrying it a great deal of fragmented stone has to be removed in getting out whole stones of any size, cast stone can be economically produced as a substitute for natural stone. Cast stone is made from a mixture of crushed natural stone and cement, the wet mix being cast in moulds. Cast stone is extensively used in lieu of natural stone.

### Natural Stone

The natural stones used in building can be classified by reference to their origin as:

(1) Igneous; (2) Sedimentary and (3) Metamorphic.

(1) The igneous stone principally used in building is granite, which was formed from the fusion of minerals under great heat below the earths surface many thousands of years ago. As the earth gradually cooled, its surface shrank and folded and parts of the layer of igneous material were forced towards the surface. The now solidified mass of igneous rock in the form of mountains and rocky outcrop is generally known as granite. It consists principally of quartz, felspar and mica. The hardest and most durable stone used in building is granite.

(2) Sedimentary stone was formed gradually over thousands of years from particles of calcium carbonate or sand deposited by settlement in bodies of water. Gradually layer upon layer of particles of lime or sand settled into depressions in the earth's surface and in course of time these layers of lime or sand particles became compacted by the water or earth above them. The more compact of these layers are quarried and used as limestone or sandstone. Because of the way in which it was formed sedimentary stone is stratified, that is, consists of layers of material. Because of the strata in the stone it is easier to split and cut sedimentary stones than hard igneous stones, which are not stratified. The strata also affect the way in which the stone can be used, if it is to be durable.

(3) Metamorphic stones are those that have been changed from igneous or sedimentary stone or from earth into metamorphic stone by pressure, or heat, or both in the earth's crust. Examples are marble which was formed from limestone and slate and shale formed from clay.

The classification of stones by reference to the nature of their origin is only a general classification. A more particular classification by use is generally used today and building stones are grouped as:

(1) Granite and other igneous stones.
(2) Limestones and marble.
(3) Sandstone.

### Granite

Granite consists of grains of quartz in combination with felspar and mica. Quartz is similar in composition to sand and forms the lustrous, colourless particles in granite. Felspar crystals contain lime and soda with other minerals in varying proportions. The colour of a granite and its hardness and durability is determined by the composition of the felspar in it. A hard durable granite contains a preponderance of quartz grains with bright hard grains of crystaline felspar, and its surface looks bright and lustrous. A poor granite containing soft mineral particles has a dull looking surface. The granites principally used are:

**Aberdeen granite** which is quarried from deep beds of rock around Aberdeen. Much of the granite in these beds is of a uniform density and composition and free from faults (cracks) and can be quarried and cut into comparatively large stones. The best known Aberdeen granites are Peterhead granite, which is pink, Rubislaw, which is blue grey and Kemnay which is grey. All of these granites are fine grained and hard and can be finished to a smooth polished surface. Aberdeen granite is generally used as a facing material for larger buildings and for monumental work.

**Devon and Cornish granites** are coarse grained, light grey in colour with pronounced grains of white and black crystals visible. The stone is very hard and practically indestructible. Because it is coarse grained and hard it is laborious to cut and shape and cannot easily be finished with a fine, smooth finish. These granites have been principally used in engineering works for bridges, lighthouses and docks. They have also been used in the walls of monumental buildings and as a walling material in the counties of their origin. Granite is at present being quarried in Penryn (Cornwall), De Lank (Cornwall) and Penzance (Cornwall).

The igneous stones, other than granite, used in building were formed thousands of years ago from the eruptions of volcanoes from which molten lava, dust and mud were ejected and settled and consolidated into layers of hard stone. This type of igneous stone is particularly hard and laborious to cut and shape. In the mountain regions of Cumberland and Scotland it is used for rubble walling but is little used elsewhere for building.

**Limestones**

Limestones were formed from the shells and skeletal remains of simple marine or fresh water organisms. Over thousands of years the shells of these organisms settled to the bed of sea or lake and gradually became consolidated into rock strata. The limestones used in building consist principally of grains of calcium carbonate bound together with the same material.

Limestone is a stratified rock but it is difficult to distinguish the strata, or layers, by eye in most hard durable limestones. Most limestones are fine grained and can be comparatively easily cut and shaped. The limestones principally used are:

**Portland Stone** is quarried in Portland Island on the coast of Dorsetshire. There are extensive beds of this stone which is creamy white in colour, weathers well and is particularly popular for facing larger buildings in towns. Many large buildings have been built of Portland stone because an adequate supply of large stones can be obtained from the quarries, the stone is fine grained, delicate mouldings can be cut on it and it weathers well even in industrial towns. Among the buildings constructed with this stone are the great Banquetting Hall in Whitehall (1639), St. Paul's Cathedral (1676) the British Museum (1753), Somerset House (1776). More recently, large buildings have been faced with this stone.

In the Portland stone quarries are three distinct beds of the stone, the base bed, the whit bed and the roach. The base bed is a fine, even grained stone which is used for both external and internal work to be finished with delicate mouldings and enrichment.

The whit bed is a hard, fairly fine grained stone which weathers particularly well, even in towns whose atmosphere is heavily polluted with soot, and it is extensively used as a facing material for large buildings. The roach is a tough coarse grained stone which has principally been used for marine construction such as piers and lighthouses.

The stones from the different beds of Portland limestone look alike to the layman. It is sometimes difficult for even the trained stonemason to distinguish base bed from whit bed. Roach may be distinguished by its coarse grain and by the remains of fossil shells embedded in it.

When taken from the quarry the stone is moist and comparatively soft, but gradually hardens as moisture (quarry sap) dries out.

The Portland stone quarries are extensively worked today and supply considerable quantities of stone for facing buildings and for ornamental stone work. A result of the quarrying operations is that a large quantity of fragmented limestone is left after large stones have been cut. This small broken stone is of no use in building but increasingly during recent years has been used in the production of reconstructed cast stone.

**Bath stones.** Many of the buildings in the town of Bath were built with a limestone quarried around the town. This limestone is one of the great Oolites and a similar stone was also quarried in Oxfordshire. Bath stone from the Taynton (Oxfordshire) quarry was extensively used in the construction of the early colleges in Oxford (St. Johns for example) during the 12th, 13th and 14th centuries. Many of the permanent buildings in Wiltshire and Oxfordshire were built of this stone which varies from fine grained to coarse grained in texture, and light cream to buff in colour. Most of the original quarries are no longer worked because it is expensive to get out stones of any size and there is not sufficient demand for the stone at present to warrant opening up the old quarries.

The durability of Bath stone varies considerably. Some early buildings built with this stone are well preserved to this day but others have so decayed over the years and been so extensively repaired that little of the original stone remains. Extensive repair of the Bath stone fabric of several of the colleges in Oxford is being carried out and further repair is necessary.

Bath stone is not extensively quarried today. Two of the Bath stone quarries being worked are:

**Corsham Down (Wiltshire).** Produces a fine grained, dark cream coloured stone which weathers well and is used for masonry walling and as a stone facing.

**Box Ground (Wiltshire).** Produces a coarse grained, cream coloured stone which weathers well and is quite

extensively used for masonry walling and as a stone facing.
Bath stone is quite soft and moist when first quarried but hardens gradually.

Other limestones being quarried for building are:

**Ancaster stone.** Quarried at Ancaster in Lincolnshire· Two qualities of this stone are used. The fine grained freebed which is cream in colour, weathers well and can be used for masonry walls, stone facing or internal work. The coarse grained shelly weatherbed is brown in colour and unsuitable for use externally.

**Beer stone.** Quarried near Seaton in Devonshire. It is a soft, white, fine grained limestone which is easy to work and can be used mostly for interior work. It is particularly suitable for carving and has been considerably used in churches. Because of its fine colour and ease of working it is being increasingly used.

**Doulting.** Quarried at Doulting in Somerset. It is a coarse grained, cream coloured stone suitable for exterior or interior work. It is quite extensively used today, principally for interior work.

**Kentish Rag.** A compact stone consisting mainly of crystalline limestone but containing also grains of quartz. It is quarried at Maidstone in Kent.

Two of the quarries which have in the past produced a limestone extensively used in building, but which are little worked, are Hopton Wood and Clipsham.

**Sandstone**
Sandstone was formed from particles of rock broken up over thousands of years by the action of winds and rain. The particles were washed into and settled to the beds of lakes and seas in combination with clay, lime and magnesia, and were gradually compressed into strata of sandstone rock. The particles of sandstone are practically indestructible and the hardness and resistance to the weather of this stone depends on the composition of the minerals binding the particles of sand. If the sand particles are bound with lime the stone often does not weather well, as the soluble lime dissolves and the stone disintegrates. The material binding the sand particles should be insoluble and crystalline. The durability of a particular sandstone in an industrial atmosphere may be tested by immersing a sample of it in an acid solution to determine the solubility of the materials cementing the sand particles. Sandstones are coarse grained and cannot as easily be worked to delicate mouldings as can fine grained limestones.
The stratification of most sandstone is visible even to the layman's eye, as fairly close spaced divisions in the sandy mass of the stone. It is essential that this type of stone be laid on its natural bed in walls.
Many sandstones are quarried in the northern counties of England where for centuries the stone has been the material commonly used for the walls of permanent buildings.
Of recent years sandstone has been increasingly used particularly in the northern counties of England and in Scotland. The quarrying and cutting of the stone has been mechanised in several quarries, which can now supply quantities of good building sandstone.

Some of the sandstones used today are:

**Crosland Hill (Yorkshire).** A hard light brown sandstone of great strength which weathers well and is used for masonry walls, as a facing and for engineering works. It is one of the stones known as hard York stone, a general term used to embrace any hard sandstone not necessarily quarried in Yorkshire.

**Blaxter stone (Northumberland).** A hard, cream coloured sandstone which is being quite extensively quarried. Used for masonry walling and as a facing material.

**Doddington (Northumberland).** A hard, pink sandstone used for masonry walling in the northern counties of England and in Scotland.

**Darley Dale (Derbyshire).** A hard durable sandstone of great strength, much used for engineering works and also for walling in buildings. It is hard to work and generally used in plain, unornamented walls. Grey, pink and brown varieties of this stone are quarried.

**Crow Hall (Durham).** A hard, light brown stone which weathers well and is used for masonry walling.

**Forest of Dean (Gloucestershire).** A hard, durable, grey or blue-grey sandstone which is hard to work but weathers well and is used for masonry walling.

**Durability of natural stone**
Natural stone has been used in the construction of the walls of larger buildings because it was thought that any hard natural stone would resist the action of wind, rain and frost for centuries. In the event there have been some notable failures of natural stone used in the construction of important buildings in this country and such failures, coupled with the cost of the material, have greatly reduced the popularity of stone as a building material.
The best known instance of decay in natural building stone occurred in the fabric of the Houses of Parliament, the walls of which were built with a magnesian limestone from Anston in Yorkshire.
A Royal Commission reported in 1839 that the magnesian limestone quarried at Bolsover Moor in Yorkshire was considered the most durable stone to be obtained for the fabric of the Houses of Parliament. After building had commenced it was discovered that the Bolsover Moor quarries were unable to supply sufficient large stones for the building and a similar stone from the neighbouring quarry of Anston was

chosen as a substitute. The quarrying, cutting and use of the stone was not supervised closely and in consequence many inferior stones found their way into the building and many otherwise sound stones were incorrectly laid. Decay of the fabric has been continuous since the Houses of Parliament were first completed and extensive, costly renewal of stones has been going on for many years. At about the same time that the Houses of Parliament were built, the Museum of Practical Geology was built in London of Anston stone from the same quarry which supplied the stones for the Houses of Parliament, but the quarrying, cutting and use of the stones was closely supervised for the Museum, whose fabric remained sound.

The above instances of misuse and correct use of the same natural stone point to another factor that has caused stone to be less and less used. A chemical analysis of stone does not entirely indicate its suitability as a building material. Two stones of similar chemical composition and from the same quarry will often resist the action of weather quite differently. The selection of sound building stone is largely a matter of very considerable experience in the handling and use of the material. It is obviously impossible for most builders, architects and engineers to gain this experience unless they deal entirely with masonry buildings and they have, over the years, become wary of using stone because of the known failures that have occurred.

But sound natural stone can be a beautiful, immensely durable material for buildings. Selected with care and used with feeling and artistry by craftsmen, natural stone is the very foundation and being of most of the buildings which are a monument to builders throughout the ages.

The durability of sound Aberdeen, Devon and Cornish granite can only be measured in centuries. But the very hardness of these stones limits their use in building because of the great cost of quarrying and cutting them.

Of the limestones, Portland stone is undoubtedly the most highly regarded. The quarries can supply adequate stones of uniform quality for all building needs. The stone weathers extremely well and is fine grained and can readily be cut and moulded. It can be used for solid masonry or as a facing and even in atmosphere heavily charged with smoke it weathers well. Hard sandstones, such as York stone, weather extremely well in all but heavily polluted atmospheres where the coarse texture of the stone soon becomes soot encrusted and may in time flake on the surface. Sandstone is hard to work and is not readily cut and moulded and does not at present enjoy the popularity of Portland limestone outside the counties in which it is quarried.

**Seasoning natural stone**

All natural stone is comparatively soft and moist when first quarried but it gradually hardens. All building stones should be seasoned (allowed to harden) for periods of up to a few years, depending on the size of the stones. If unseasoned stones are used in walls frost may cause moisture in them to expand so fiercely that the stone disintegrates. Once stone has been seasoned it does not revert to its original soft moist state on exposure to rain, but on the contrary hardens with age.

**Bedding stones**

It is important that sedimentary stones (limestones and sandstones) be built into walls so that they lie on their natural beds. The bed of a stone is its face parallel to the strata (layers) of the stone in the quarry. When stones are cut, care is taken to ensure that their strata will be at right angles to the stress they will have to suffer when used in a building. In walls the compressive stress is vertical and the bed (strata) of the stone is arranged so that it is horizontal. A stone which is not built in on bed may split along one or more strata, due to the weight of the wall above. The stratification of limestones is often very difficult to determine and it is left to the quarrymen to cut the stones so that they will lie on bed in walls.

## RUBBLE WALLING

Sound building stone is often hard and laborious to cut and shape and for centuries walls have been built with stones of irregular shape taken from the surface of land or collected from the rubble produced by quarrying large stones. The size and shape of rubble stones varies enormously, from the roughly square rubble of sandstone to the very irregular rubble of granite. The following is a description of the types of rubble walling used.

**Random rubble—Uncoursed.** The wall is built of stones of random shape and size selected to ensure a vertical stable wall and to avoid cutting of stones. The only shaping of stones that is executed is the removal of inconvenient corners or projections with a walling hammer.

Fig. 1 is an illustration of this type of wall. It will be seen that stones of different shapes and sizes are used and laid without regular horizontal courses and that the stones are only bonded by virtue of their random shapes. In general stones are laid with their long axis roughly horizontal and along the length of the wall. As the wall is built selected long stones are laid to bond into the thickness of the wall, one bonder to about every square metre of wall surface. These bonders are chosen so as to be about 2/3rds of the wall thickness in length. Bonders built right across the thickness of the wall should be avoided as they may allow rain to penetrate the mortar joint running through the thickness of the wall.

It will be seen from Fig. 1 that large roughly square stones are built in at quoins. These stones are selected

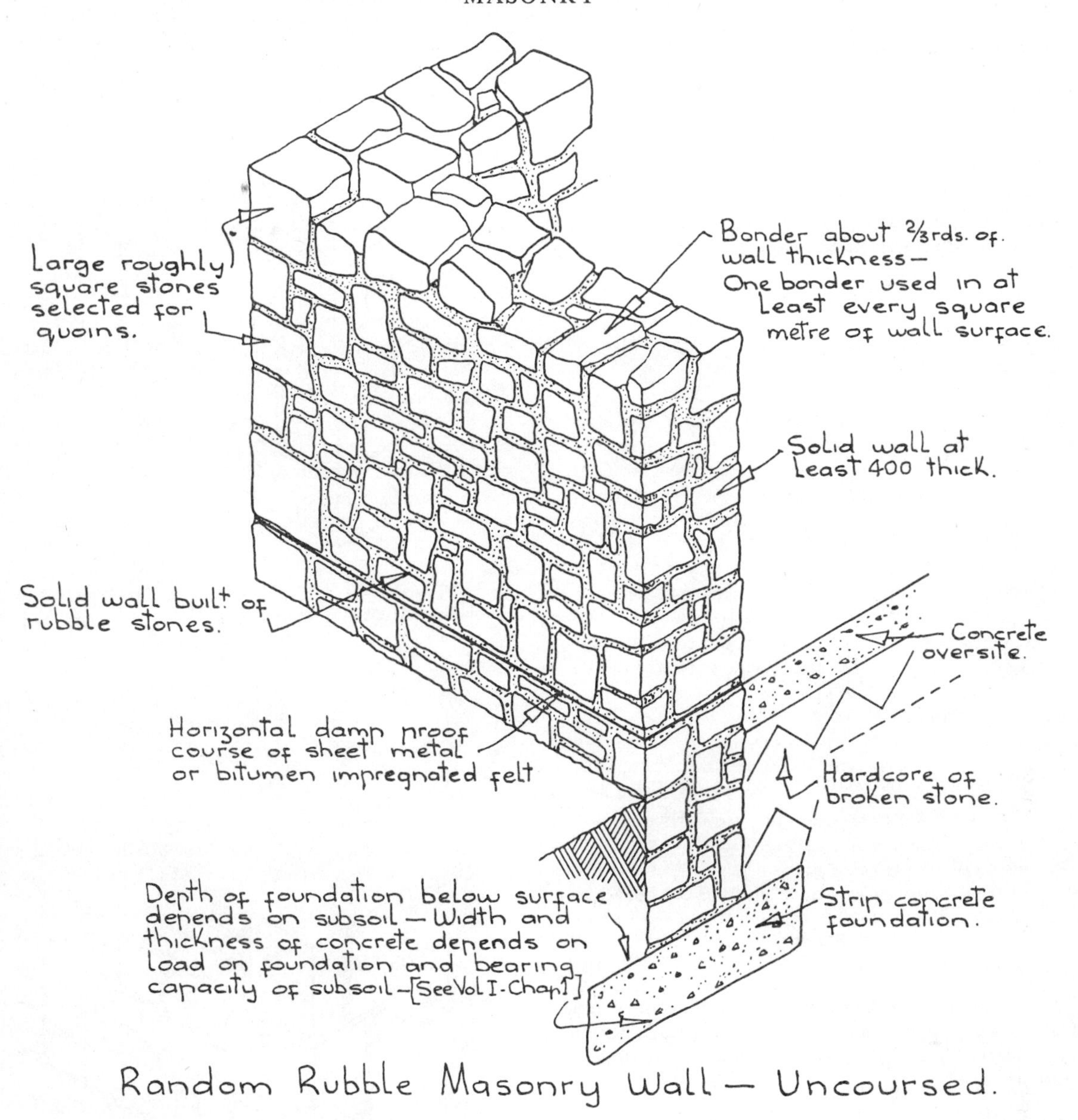


**Fig. 1**

for their size and shape and if necessary are roughly squared with the walling hammer. Their purpose is to improve the stability of the rubble wall at quoins and to provide a neater finish than is possible with ordinary random rubble. Usually the wall is built on a foundation of large roughly square stones as illustrated in Fig. 1.

**Random rubble—Brought to courses**

The rubble of sedimentary rocks often consists of roughly square stones which can often be put together in rough horizontal courses. By building a wall in courses its stability is improved. Fig. 2 is an illustration of this type of wall.

Again large roughly square stones are built in at quoins and the rubble is built so that it roughly courses in with the quoin stones. The height of the courses obviously varies with the type of stone used in the walling and is anything from say 300 to 900.

The stones of random rubble walls for buildings are usually laid in mortar. The mortar used today for this type of walling consists of cement, lime and sand in the proportions 1 - 2 - 9.

Because of the irregularity of the stones used, the mortar joints in this type of wall are often quite thick and in time the mortar may crack and disintegrate. This in turn may allow one or more stones to sink, which in turn may cause further cracking of mortar and so on. The life of a random rubble wall depends on the skill of the mason in choosing and placing stones and the degree to which the wall is exposed to rain, wind and frost.

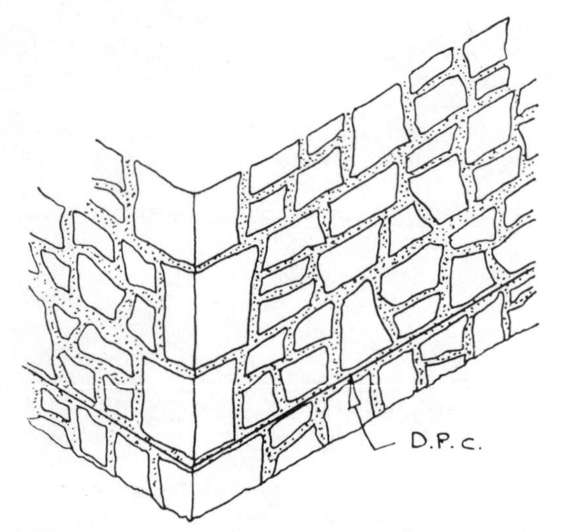


Random Rubble Masonry Wall – Brought to courses:
Stone rubble is laid so as to course in with quoin stones
Course heights vary from 300 to 900.

**Fig. 2**

The British Standard Code of Practice, Masonry rubble walls, recommends that random rubble walls should be at least 400 thick to resist rain penetration and for stability.

The stability and life of a rubble wall can be much improved if stones are first roughly squared before they are used. The stones can be bonded more accurately and thinner mortar joints used than is practicable with random rubble walling.

**Squared rubble—Uncoursed.** The stones are roughly squared and built without continuous horizontal courses as illustrated in Fig. 3. Large quoin stones are used for the reasons previously explained.

**Squared rubble—Brought to courses**
The stones are roughly squared and laid in courses to bond in with the larger quoin stone as illustrated in Fig. 4.

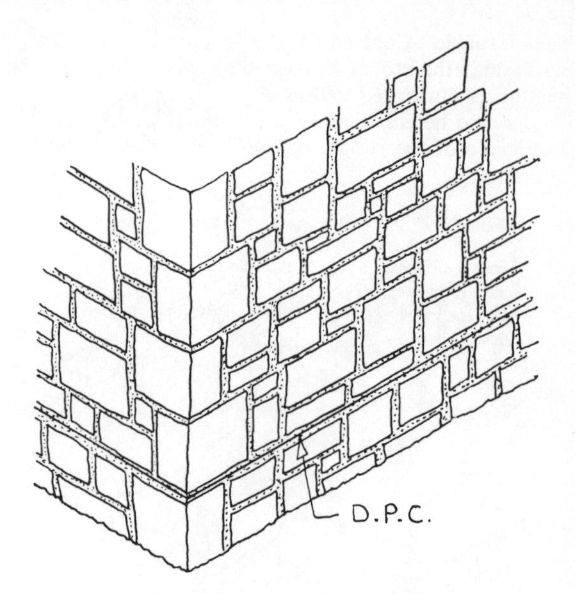


Squared Rubble – Uncoursed.
The rubble stones are roughly squared and laid without continuous horizontal courses.

**Fig. 3**

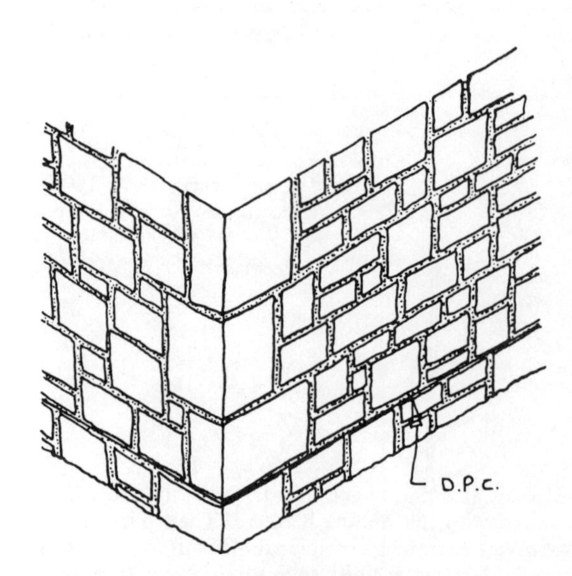


Squared Rubble–Brought to Courses:
The rubble stones are roughly squared and laid to course in with quoin stones– Course height from 300 to 900.

**Fig. 4**

**Square rubble—Coursed.** The stones are cut and squared so that the stones in each course are of the same height, as illustrated in Fig. 5.
The course heights will vary depending on the nature and shape of the rubble before it was squared.

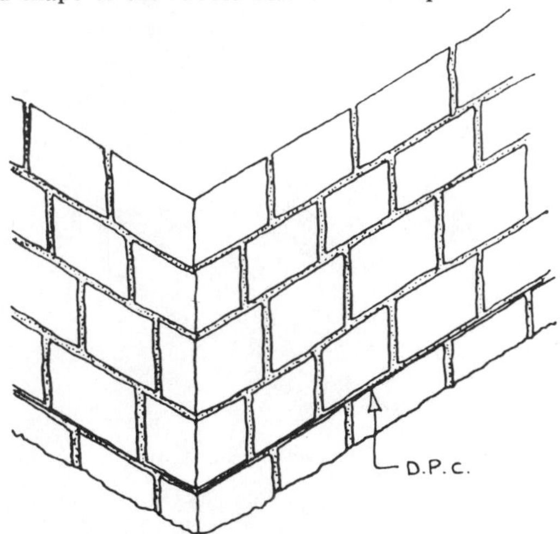
D.P.C.
Squared Rubble — Coursed.
The rubble stones are cut and squared so that they can be laid in horizontal courses — Course heights vary from 75 to 450 Average course height 225

Fig. 5

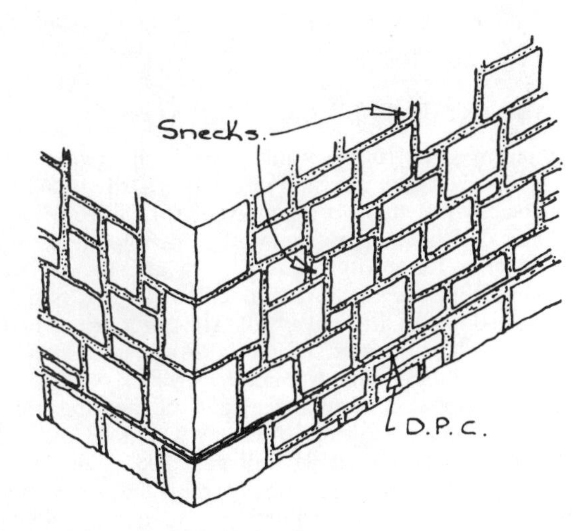
Snecks.
D.P.C.
Snecked Rubble:
The rubble stones are roughly squared and laid without continuous horizontal courses — Snecks are small squared stones, not less than 75 in any dimension.

Fig. 6

Which of the three types of squared rubble illustrated is used will depend on the ease with which the rubble can be squared, whether the stones can be matched for height without too much selection, the skill of the mason and the cost of the work.

**Snecked rubble.** This is virtually indistinguishable from squared rubble—uncoursed. The stones are roughly squared and built without continuous courses. The name sneck derives from small squared stones specially cut to fill gaps that would otherwise occur between the larger squared rubble in the wall. Fig. 6 is an illustration of this type of walling.
If the illustration of squared rubble—uncoursed (Fig. 3) is compared to the illustration of snecked rubble, (Fig. 6) the absurdity of giving different names to similar types of walling will be appreciated. It is to be hoped that the use of the description snecked rubble will be discontinued in time.

The British Standard Code of Practice, recommends that squared rubble walls should be at least 400 thick. In practice squared rubble walls 300 thick have resisted penetration of rain and have remained stable. Obviously the least thickness of a wall depends on the size of stones used and the ease with which they can be bonded without cutting.

**Polygonal rubble walling. (Kentish Rag).** The rubble stones of rocks which are not obviously stratified and are hard and laborious to dress to roughly square shapes, are often roughly shaped and built as polygonal walling. Fig. 7 is an illustration of this type of walling. It will be seen that the joints run irregularly and follow the odd shapes of the stones. The limestone known as Kentish Ragstone was commonly used in this way.

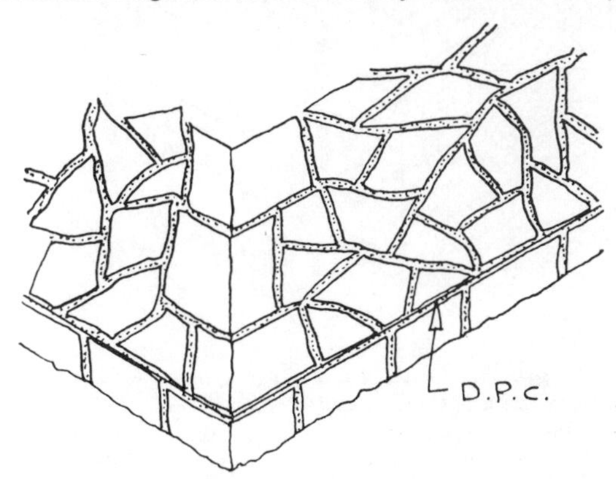
D.P.C.
Polygonal rubble walling.
The rubble stones are cut only as necessary to fit them together with irregular mortar joints as shown. Squared stones or brickwork used at quoins, jambs and in foundations.

Fig. 7

Obviously the strength of this walling depends on the skill and patience of thè mason in shaping and fitting the stones together.
The mortar used for squared rubble walling is the same as that described for random rubble.

**Flint walling.** In the eastern and south eastern counties of England there are no extensive beds of sound building stone. But in and around the chalk hills and on the sea shore are numerous flints (cobbles). Flints are small irregularly shaped lumps of nearly pure silica varying in size from 75 to 150 in width and up to 300 long. The core of these flints is steel grey in colour and their natural surface colour is white. Flints are very hard, but being brittle can easily be split. On the east and south east coast of England, rounded sea shore flints were commonly used as a facing for rubble walls of buildings. Selected flint pebbles of roughly the same size were used as a facing, with flints of irregular shape in the heart of the wall built in lime mortar. The quoins and jambs of these walls were built with squared natural stone or brick as illustrated in Fig. 8. Walls built of flint are not particularly stable because of the small size of the material used and it was usual to build lacing (bond) courses of bricks or stone at intervals as illustrated in Fig. 8. The base of flint walls was usually constructed with one or more courses of squared stones or a few courses of brickwork.

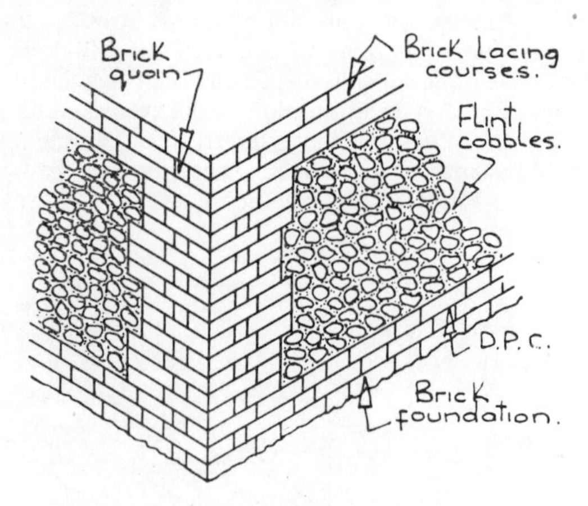


Fig. 8

Inland flints were sometimes split (knapped) roughly square and built in courses as illustrated in Fig. 9. Flint walling is very rarely built today because the operation of knapping (snapping) the flints is costly.

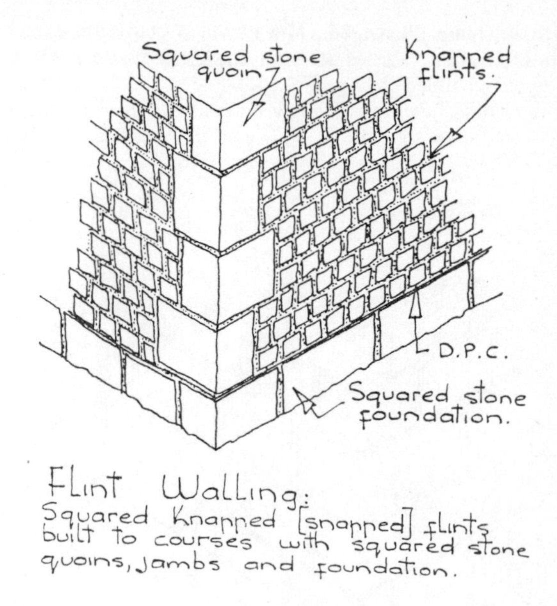


Fig. 9

With the mass production of bricks that followed the industrial revolution in this country walls of natural stone rubble were less frequently built, other than in districts distant from road or rail transport. A rubble wall may have to be at least 400 thick to resist penetration by rain, it is a poor insulator against loss or gain of heat and is more expensive to build than a brick wall.

## MASONRY RUBBLE CAVITY WALLS

Of recent years rubble walling has been quite extensively used as the outer skin of cavity walls particularly in areas where there is a supply of rubble from local quarries. Rubble masonry walling is chosen as the outer skin for cavity walls for its appearance, the variations in size, texture and colour of the natural stone making an attractive wall. The outer skin is built of random or squared rubble with an inner skin of bricks or blocks and a cavity 75 wide. A random rubble outer skin should generally be not less than 400 thick because a lesser thickness would be difficult to bond to form a stable wall and also might allow rain to penetrate it so heavily that the inner skin became damp. The thickness of a squared rubble outer skin will depend on the density of the stone used and the ease with which it can be squared, a thickness of 225 to 300 being usual. A 75 cavity is considered necessary in this type of wall to ensure that the air in it does not become heavily moisture laden and to allow sufficient space for the mason to bed and bond the stones and at the same time to keep the cavity clear of mortar droppings and stone debris.

Fig. 10 is an illustration of a cavity wall with squared rubble outer skin and concrete block inner skin. The construction around window openings and the support of floors and roof is shown.

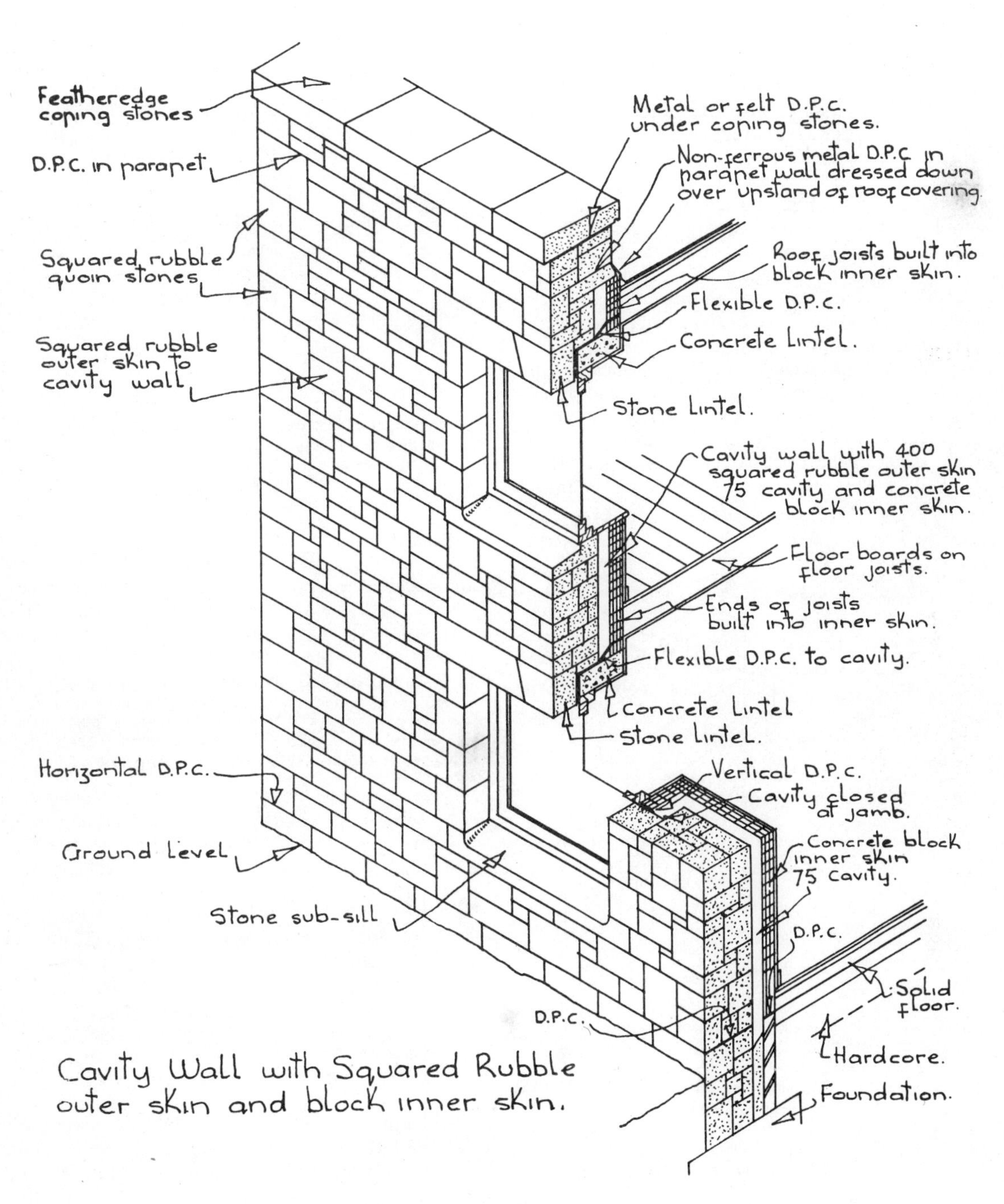


**Fig. 10**

## ASHLAR MASONRY WALLING

The words ashlar masonry describe the use of stones very accurately cut and finished true square to specified dimensions so that they can be built with very thin mortar joints. Obviously the labour involved in cutting and finishing ashlar stones is such that ashlar masonry is a very expensive form of wall, and today it is used mainly for larger permanent buildings. Because ashlared stones are expensive they are commonly bonded to a brick backing in walls, the stones being cut to correspond to brick course heights and to bond in with brick widths. Fig. 11 is an illustration of part of a wall faced with ashlar masonry bonded to brick backing. Ashlar masonry may be bonded to a brick backing in solid walling or bonded to brick backing as the outer skin of a cavity wall with an inner skin of brick or blocks. The least thickness of solid ashlar faced wall to resist rain penetration in all but sheltered positions is 1½B. This thickness includes ½B and 1B thickness of ashlar in alternate courses and 1B and ½B of brick backing. In exposed positions the thickness of a solid ashlar faced wall should be 2B, the additional ½B thickness being in the brick backing. In ashlar faced cavity walling the outer skin is usually 1B thick with stones 1B and ½B thick in alternate courses. Figs. 11 and 12 illustrate typical solid and cavity walls faced with ashlar. 1B indicates the length of a standard brick (see Vol. 1).

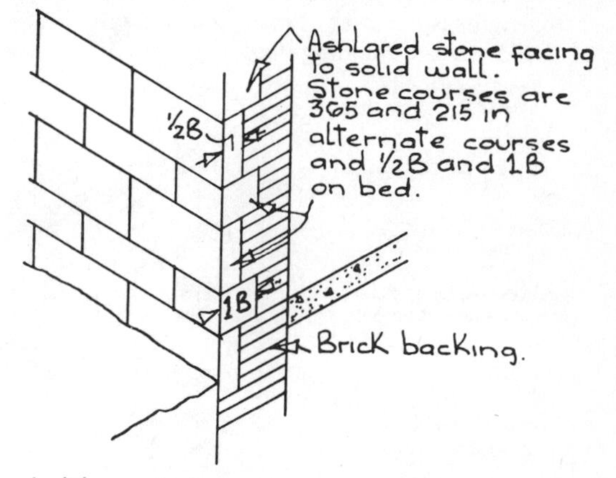


Fig. 13

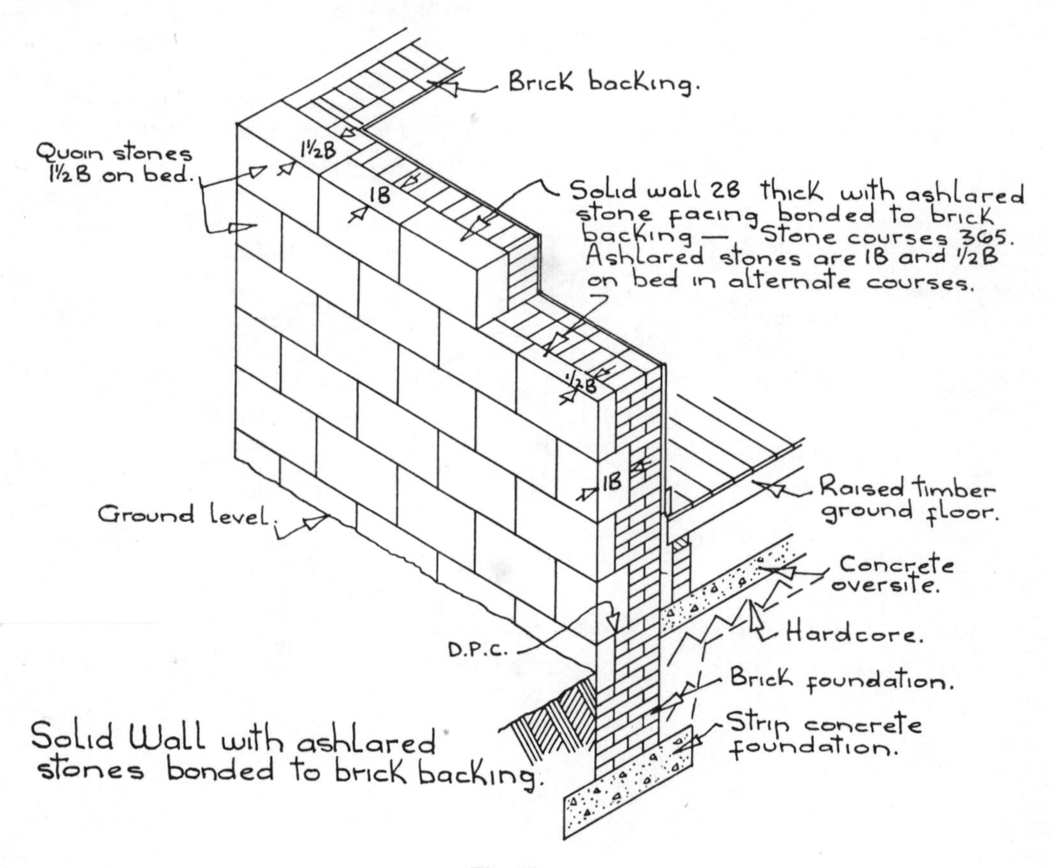


Fig. 11

**Size of ashlar stones.** It will be seen from Fig. 11 that the ashlar stones are of differing thickness in succeeding courses, the difference being ½B to bond in with the brickwork backing. The height of the stone courses is a multiple of brick course heights. The choice of course heights is a matter of taste, the general rule being that the larger the building the larger the stones used. The ashlar stone courses are not necessarily of the same height throughout the wall. Alternate courses may be of different heights so that if the courses of thicker stones are of less height than the courses of thinner stones, there is some saving in the amount of stone used. This is illustrated in Fig. 13. The least thickness of stone that can successfully be used is

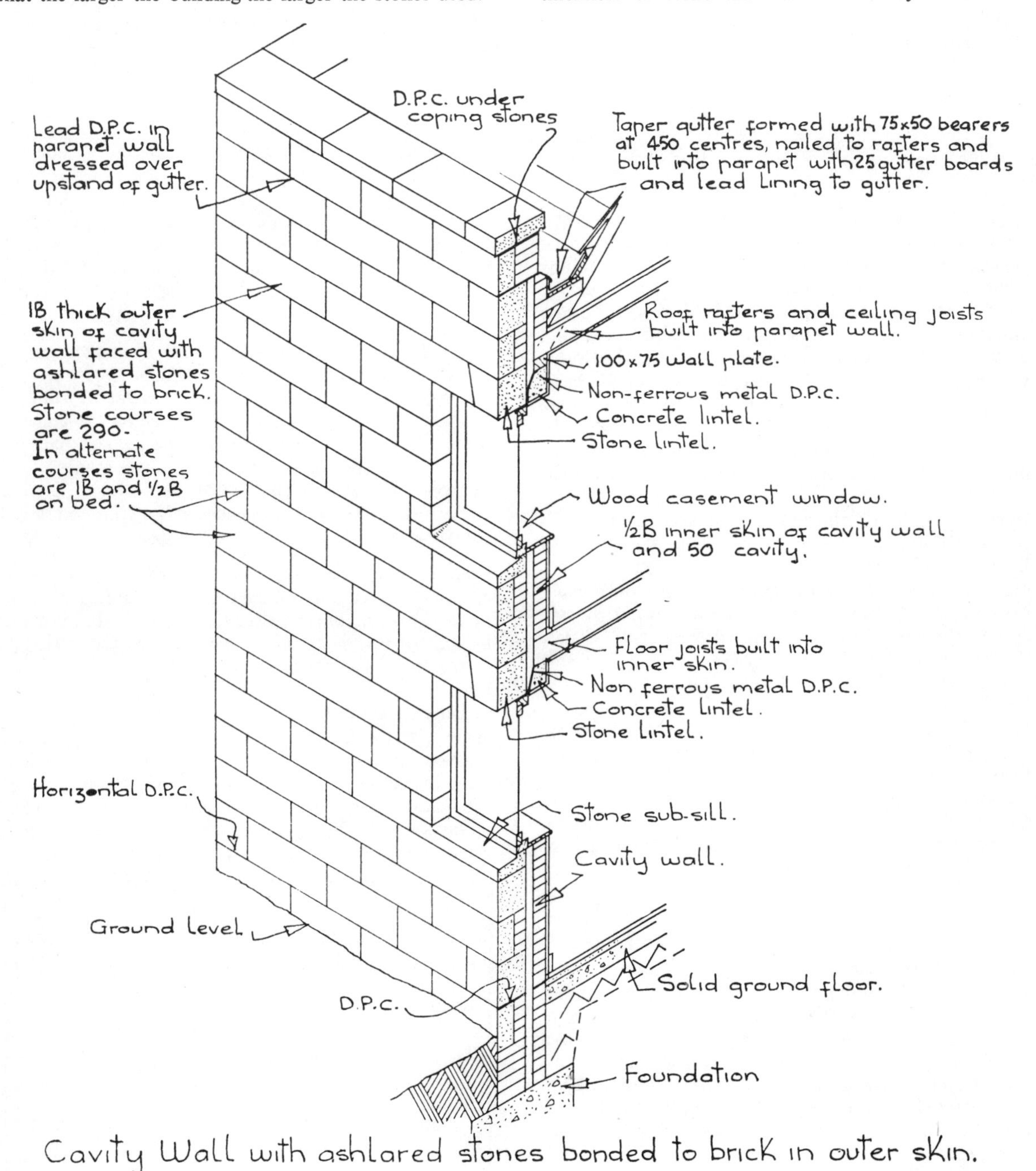


Cavity Wall with ashlared stones bonded to brick in outer skin.

**Fig. 12**

generally considered to be 100 for stones such as Portland limestones which does not have pronounced stratification and 115 or more for stones with pronounced stratification such as sandstone.
The length of each stone is determined firstly by the size of the blocks of sound stone that can be cut from the quarry face and secondly by the design of the building.
As a general rule stones are cut so that their length is about 1½ to 1¾ times their height where stone courses are of regular height, or about 600 to 900 long where courses are of irregular height. Obviously as each ashlar stone is cut to a specified size to suit a particular building there are no standard lengths for stones.

**Type of stone used for ashlar masonry**. Whether granite, limestone or sandstone is used for ashlar masonry depends on the taste of the Architect and the building owner, the locality in which the building is to be erected and the estimated cost of the work. Sound granite is very hard and therefore expensive to cut and dress as ashlar stone. Because it is virtually indestructible it is often used for ashlar stones at the base of walls where other softer stone might perish in time. Limestone, particularly Portland limestone, weathers well and is comparatively easy to cut and dress and is more commonly chosen for ashlar masonry than other natural stones.
Hard sandstone is laborious to cut and dress but is commonly used for ashlar masonry in the north of England and in Scotland. The quarrying, cutting and dressing of sandstone in several quarries has been mechanised of recent years, with the result that the cost of the stone has not increased so sharply as has the cost of other building materials.
The backs of ashlar stones to be bonded to a brick backing are usually painted with bitumen. The purpose of the bitumen is to prevent stains appearing on the face of the stone due to water drying out through them from the brick backing and to prevent an efflorescence of soluble salts appearing on the face of the stones. The lighter coloured limestones and sandstones are particularly liable to staining and stones cut from them should always be painted with bitumen on their back faces and edges before being built in.

**Masonry walls—Foundations.** Before cement was manufactured masonry walls were built on a foundation of large stones laid on firm subsoil. Today a concrete strip foundation is usual for both rubble and ashlar masonry walls. The width and thickness of the concrete depends on the loads on the foundation and the safe bearing capacity of the subsoil as explained in Vol. 1, Chapter 1. Similarly the depth of the foundation below ground is generally at least 600 to avoid damage by frost heave, or at least 1·0 on shrinkable clay subsoils. In districts where there is a sub stratum of firm rock the foundation of walls may be near the surface on a bed of concrete spread on the roughly levelled rock.
The foundations of random rubble and squared rubble walls are constructed with large roughly squared stones laid on the concrete foundation. This at once strengthens the wall at its foundation and provides a means of forming a level bed for the D.P.C.

**Damp Proof Course.** It is essential that a continuous layer of some impermeable material be built into masonry walls above ground, where there is any likelihood of dampness rising from the ground. Any of the materials noted in Vol. 1, Chapter 2 may be used but one of the metals, such as lead or copper, in sheet form is preferred for its durability. Another reason for using sheet metal as a D.P.C. is that it is less liable to damage during building operations than, say, felt. Sheet metal is also preferred as D.P.C. because it does not cause a thick ugly joint as would asphalt, for example.

**Joggle joint:** A shallow 'V' groove is usually cut in the end faces of ashlar stones. When the stones are laid a wet mix of cement and water (grout) is run into the square hole formed by the matching 'V'-grooves in the ends of adjacent stones. The cement hardens to form an interlocking joggle between stones. The purpose of these joggles is to keep stones in their correct position and resist lateral pressure on them from the brick backing or floor or roof.

**Mortar.** The mortar for masonry walling should be reasonably plastic and mixed with just sufficient water to ensure workability. After hardening it should have roughly the same density and compressive strength as the stone used in the wall.
The proportion of matrix (lime–cement) to fine aggregate for mortar is usually 1 to 3 by volume. Fine aggregate may be either clean natural sand or clean crushed natural stone, the latter being used particularly in mortar for ashlar limestone masonry as it produces a mortar with a colour and texture near that of the stone itself. The following table sets out some usual mortar mixes for masonry walling:

| *Use* | *Cement* | *Lime* | *Fine aggregate* |
|---|---|---|---|
| Rubble or ashlar walling below D.P.C... .. | 1 | ¼ | 3 |
| Walls exposed to severe conditions .. .. | 1 | ¼ | 3 |
| All normal rubble wall construction .. .. | 1 | 2 | 9 |
| Ashlar limestone and softer sandstones .. | — | 1 | 3 |
| or | 1 | 3 | 12 |
| Ashlar hard sandstone.. | 1 | 1 | 6 |
| Ashlar granite .. | 1 | — | 3 |

### Quoins

The quoins of walls built of random rubble or squared rubble are built with selected large stones square dressed and laid to bond in with the general walling as illustrated in Figures 1 to 6.

The quoins in ashlar walling are built with stones cut to size to close the bond. As previously explained ashlar walling is commonly built as a facing bonded to brick backing with stones for example ½B and 1B thick in alternate courses. If the quoin (corner) stones of the ½B thick course of stones were only ½B thick their ½B wide faces would be exposed. This narrow width of stone appearing at an angle would look insignificant and destroy the effect of strength and permanence. Thicker quoin stones are therefore built in as illustrated in Fig. 14, and bonded to the brick backing and the stone facing.

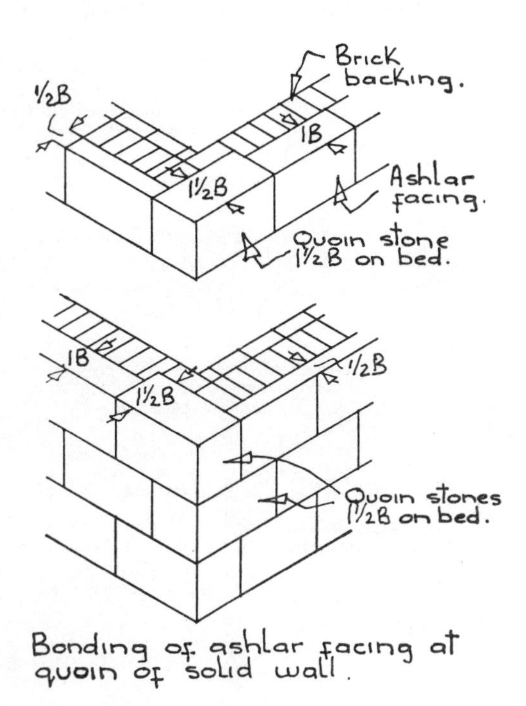


Fig. 14

## OPENINGS IN MASONRY WALLS

The names of the parts of openings for windows and doors are the same as those described for brick walls in Vol. 1, Chapter 3.

### Jambs

**Rubble masonry—solid walls.** To strengthen rubble walls and to provide a neat finish to the jambs of openings, selected squared stones are built in and bonded to the surrounding wall. As the stones in the jambs of openings have to be dressed square it is usual to finish them with a rebate behind which the wood window can be fixed. This at once affords some protection to the window frame and makes for a neat finish between frame and stone. Fig. 15 is an illustration of the jamb of a window opening in a solid squared rubble wall.

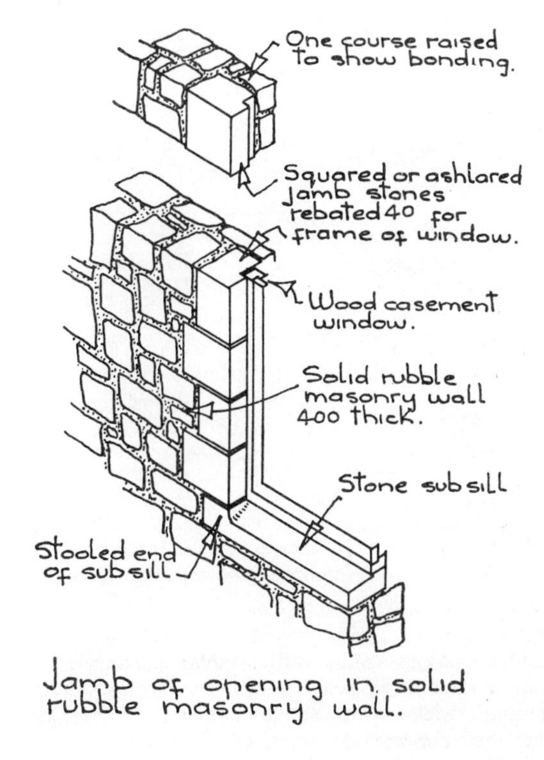


Fig. 15

The stone sill illustrated is cut from one long stone which is weathered, sunk and stooled at ends for building into jambs. If the opening is wide it is unlikely that the sill can be cut from one stone. Two or more stones are used and the joint between them on the weathered (sloping) surface of the sill is finished as a saddle joint similar to the joint used with coping stones described later in this Chapter.

**Rubble masonry—cavity walls.** As explained previously rubble walling may not resist penetration of rain, particularly in exposed positions, and it is common today for rubble masonry to be built as the outer skin of cavity walling. The cavity at the jambs should be closed for the reason given in Vol. 1, Chapter 3. If the cavity is closed with masonry or brick or blocks a vertical D.P.C. should be built in to prevent moisture penetrating the solid filling at jambs. Fig. 16 illustrates this.

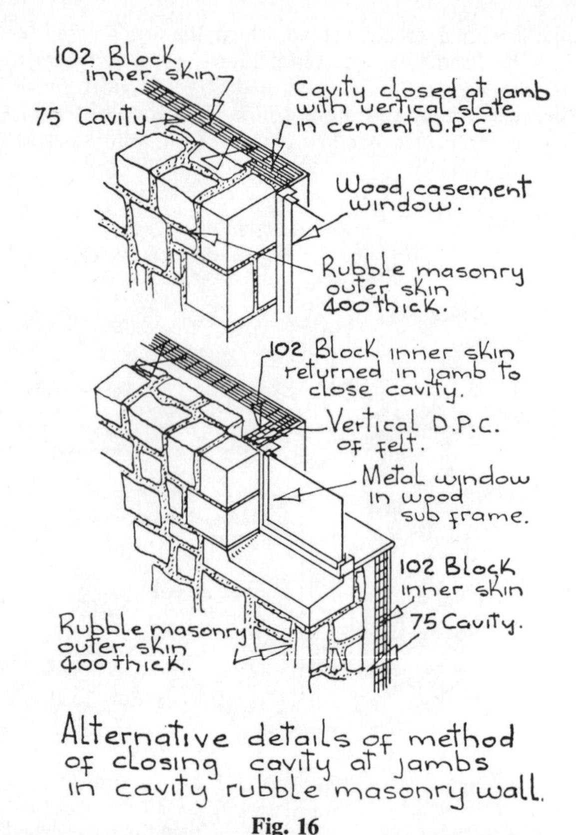


Fig. 16

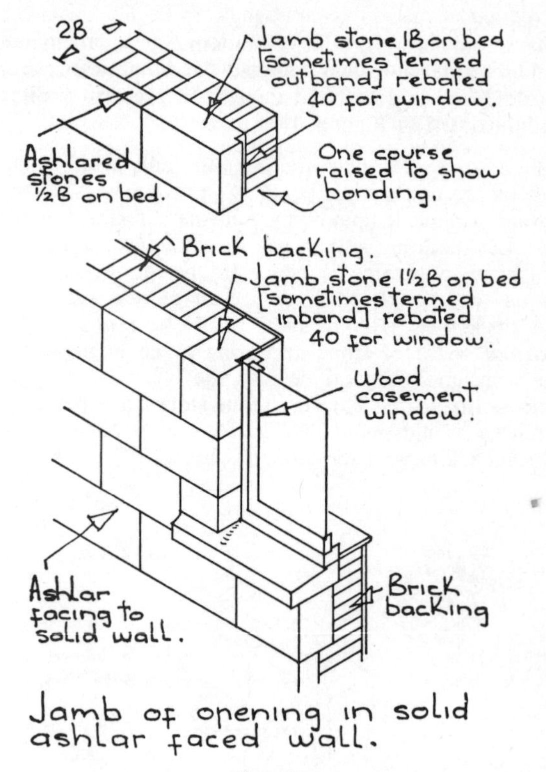


Fig. 17

**Ashlar masonry—solid walls.** Ashlar masonry is commonly bonded to brick backing with the stones of different thickness in alternate courses. To strengthen the bond between the stone and brick at jambs, and so that the ends of the thinner stones will not be exposed in the outer reveals of openings, large ashlar jamb stones are used. In alternate courses the jamb stones are cut so that they bond into the thickness of the wall (inband) and along the length of the wall (outband) as illustrated in Fig. 17.

**Ashlar masonry—cavity wall.** Fig. 18 illustrates the closing of the cavity at jambs and the vertical D.P.C. which should be built in to prevent penetration of moisture through the solid filling at jambs.

**Stone lintels**

One way of supporting the walling over the head of openings is to cut and build in one large stone. In rubble walling one large squared stone is used and in ashlar masonry one large ashlared stone. The depth of the stone used as a lintel will depend on the span of the opening and the type of stone used. A granite lintel will be stronger than one cut from one of the softer limestones. There is no rule of thumb method of determining the required depth of stone for lintels over particular widths of opening. Only an

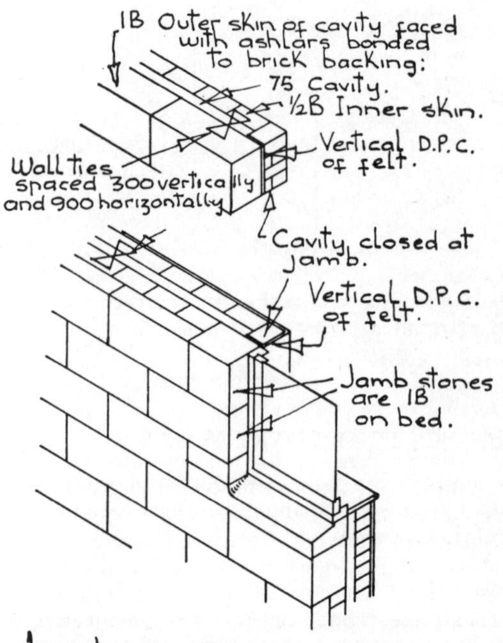


Fig. 18

experienced mason is competent to decide this from his knowledge of the behaviour of particular stones, and experience of their behaviour as lintels over many years.
The thickness of a stone lintel will depend on what size of stone can be taken out of the quarry and the position of the window or door frame in the depth of the reveal of the opening. Obviously the lintel should be at least as thick as the exposed width of outer reveal of jambs so that there is no joint in the exposed soffit of the lintel. Fig. 19 illustrates the construction of a stone lintel over a window opening.

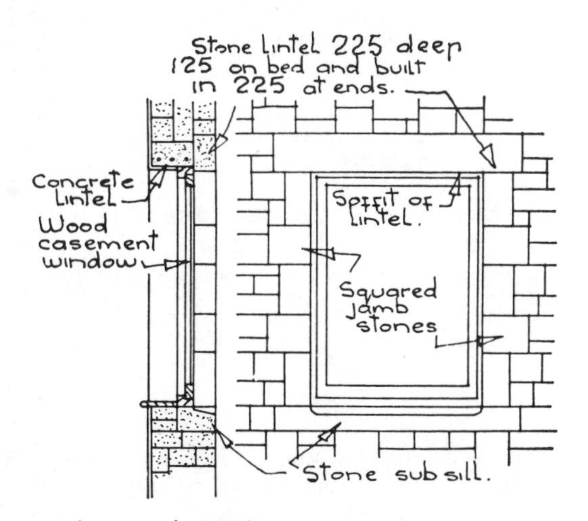


**Fig. 19**

Large stones are expensive to cut, transport and lift into position and stones sufficiently large to span openings of say more than 1·2 often cannot be taken from beds of natural stone. The usual practice therefore, is to construct lintels of three or five stones. There are various ways of building stone lintels from several stones. Fig. 20 illustrates some typical arrangements. The joints between the stones forming the lintel are cut to radiate from a centre so that they will not slip out of position under the weight of the wall above. Because the joints radiate from a centre these lintels are sometimes called flat arches.
Fig. 20 (a) illustrates one of the more usual arrangements of the stones in a flat arch or lintel. The usual arrangement is for the stones to be cut so that the length of the stones exposed on the soffit is equal and the joints between them radiate from a centre, mid-span at sill level, as illustrated in Fig. 21. The disadvantage of this type of lintel or flat arch is that due to slight settlement the centre, or key stone, may sink and spoil the appearance of the lintel. To prevent this possibility the joints between the stones are often rebated as illustrated in Fig. 20 (b). The above rebate is usually 40 deep and cut at half the depth of the lintel.

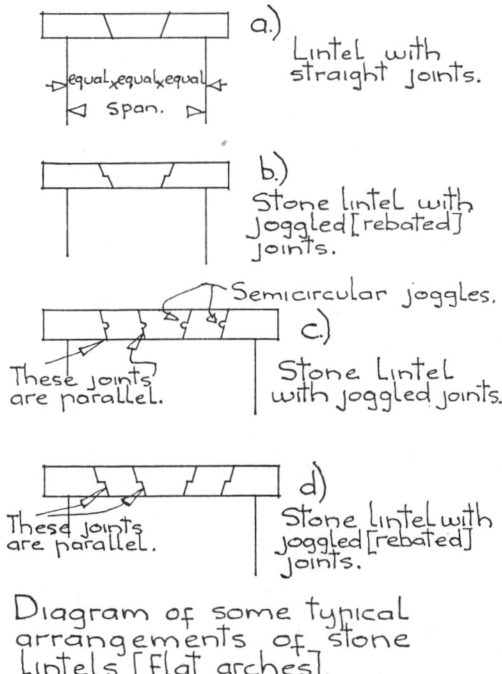


**Fig. 20**

But a rebate exposed on the face of the stones may spoil the appearance of a lintel. To avoid this a secret rebate, termed a joggle, is cut on the back of the stones. This is illustrated in Fig. 22.
In Fig. 20 (d) the lintel is constructed from five stones rebated as previously described. Alternatively the rebates may be cut as secret joggles on the inside face of the stones. In Fig. 20 (c) another arrangement is shown where five stones are used and are put together with semicircular joggles fitting to grooves in adjacent stones. These joggles serve the same purpose as rebates or secret joggles. The semi-circular joggles are usually 40 diameter, cut in the centre of the depth of the stone and right through the thickness of each stone.

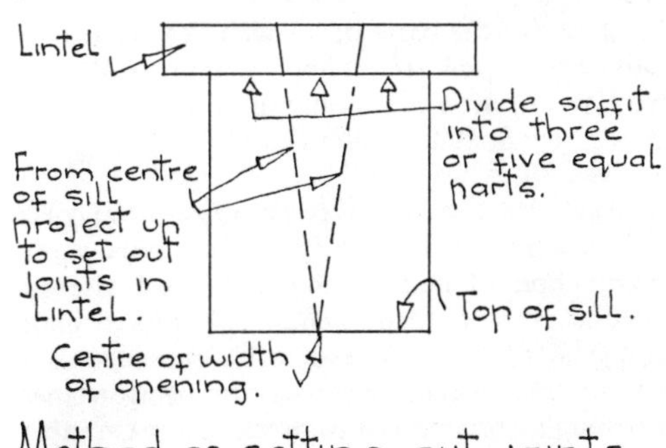


**Fig. 21**

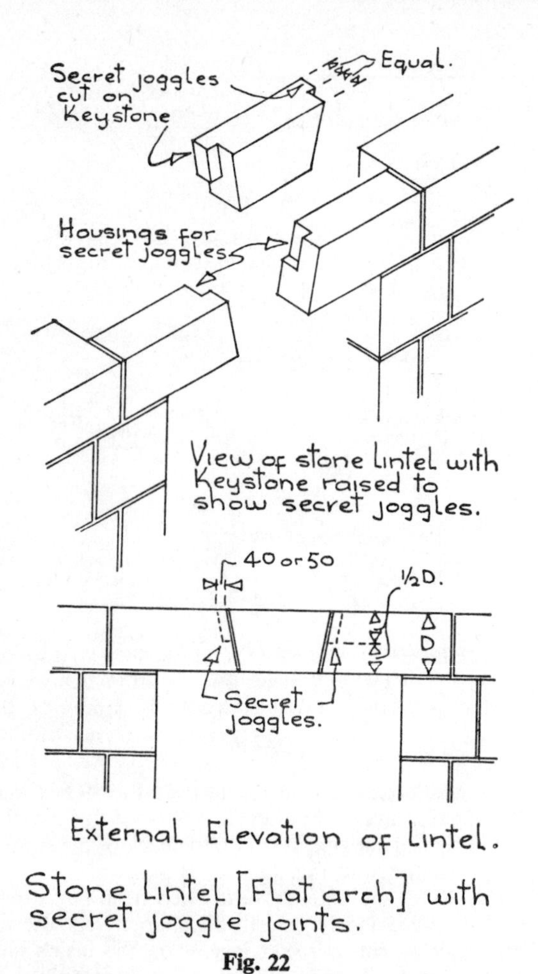


**Fig. 22**

Lintels (flat arches) in natural stone are usually 215 or 290 deep and these depths of lintel can be used to support the wall over openings of up to 1·2 width Over openings more than 1·2 wide an arch is generally used.

**Arches**

The names of the parts of an arch described in Vol. 1, Chapter 3 for brick arches apply equally to stone arches. Two types of arch commonly used in masonry walling are segmental and semi-circular arches.

**Segmental arch.** Fig. 23 is an illustration of a segmental arch over an opening in a masonry wall.
One method of determining a suitable rise for this type of arch is to calculate it as 120 for every metre of span. The particular rise of arch chosen is a matter of taste, the method given above being a means of determing a suitable rise to produce a reasonable looking arch.
Having determined the rise of the arch, the centre of the circle, of which the soffit of the arch is a segment,

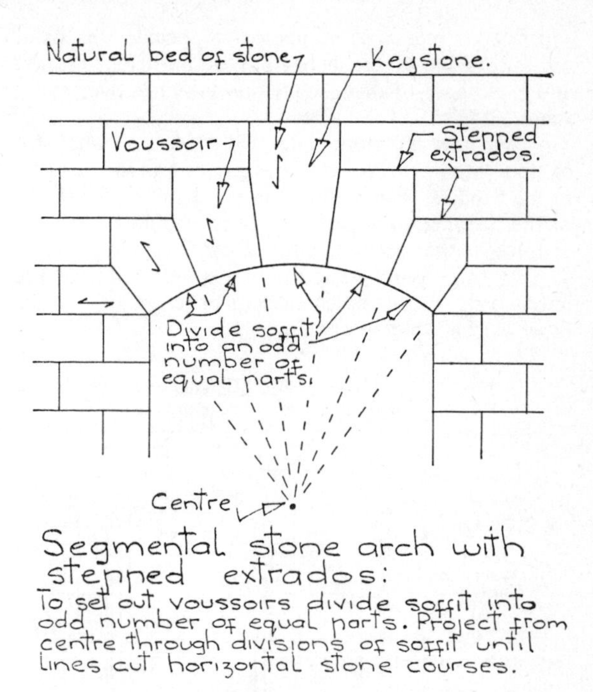


**Fig. 23**

is found as described in Vol. 1, Chapter 3. A usual way of setting out the vousoirs (arch stones) is to divide the soffit of the arch into an odd number of equal parts, each part being about two thirds of the height of the stone courses in the wall. Having drawn in the soffit of arch, the horizontal stone courses are drawn in. The arch joints are then drawn radially from the centre until they cut horizontal joints. The extrados of the arch is then drawn in, stepped from course to course towards the crown of the arch.

**Semi-circular arch.** A semi-circular arch may be set out in the same way by dividing the soffit into an odd number of equal parts and drawing the joints radially through these divisions until they strike a horizontal stone course. The stepped extrados is then drawn in. Another method of setting out an arch is illustrated in Fig. 24. It will be seen that the soffit of the arch and the stone courses around it are first drawn in. From centres on the springing line two arcs are drawn. The joints between arch stones are projected from the centre of arch to where the arcs cut horizontal stone courses. If the points A and B, in Fig. 24, are taken near the centre of the arch it will have a squat appearance and if they are near the springing points it will have a high pointed appearance.
Because the steps in the extrados of these arches are not regular, stones of irregular length have to be cut and built into the wall around the arch to complete the bond.
Instead of cutting the stones of an arch so that they

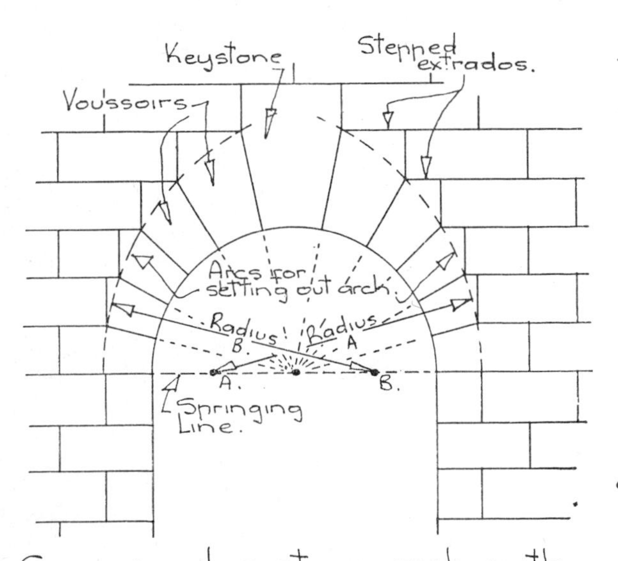


Semi-circular stone arch with stepped extrados:
To set out voussoirs choose centres A and B on springing line such that A and B are equidistant from centre of arch — with radius A equal to radius B draw setting out arcs. From centre of arch project up to points where arcs cut horizontal stone courses.

**Fig. 24**

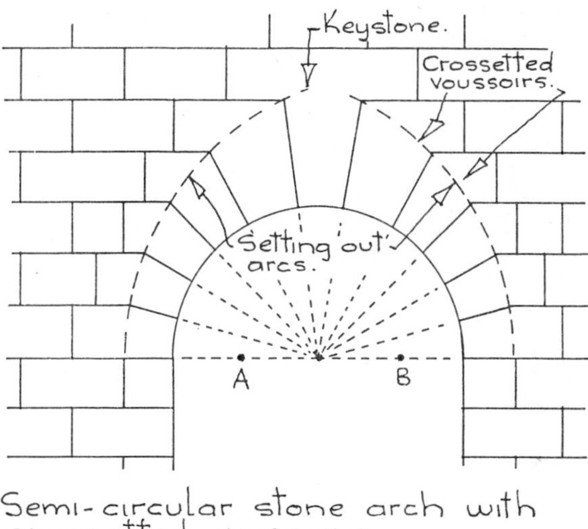


Semi-circular stone arch with crossetted voussoirs.

**Fig. 25**

bond in with the masonry around in the form of a stepped extrados, each arch stone is sometimes cut so that it radiates from the centre and is cut to bond horizontally with the masonry. It is then described as a crossetted voussoir. A semi-circular arch with crossetted voussoirs is illustrated in Fig. 25. This is an extravagant method of using stone to construct an arch and it is done purely for appearance sake. An arch with crossetted voussoirs has a massive substantial appearance and is sometimes used over the more important door openings to large buildings.

The stones of an arch are generally as thick as the exposed part of the outer reveal of openings between the face of window or door frame and the outside face of the wall. The inside thickness of the wall over the opening, behind the arch, may be supported by a rough stone arch similar to the face arch but built from roughly shaped stones. If the arch is in a wall with ashlar masonry bonded to brick backing then a rough semi-circular brick arch can conveniently be built behind the stone arch on face, or a reinforced concrete lintel with semi-circular soffit may be cast *in situ* behind the stone arch.

**Bed of stones in arches.** Stratified stones in masonry walls should be laid on their natural bed so that the strata of the stones are at right angles to the loads they support. The voussoirs in a stone arch support the weight of the wall above and transmit this weight around the curve of the arch to the jambs of the opening. It is usual to cut voussoirs from stratified stones so that the bed (strata) of the stones radiates from the centre of the arch and lies at right angles to the face of the arch as shown in Fig. 23.

## CORNICE AND PARAPET WALL

It is common practice to carry masonry walls above the level of the eaves of a roof as a parapet. The purpose of the parapet is partly to obscure the roof and also to provide a depth of wall over the top windows for the sake of the proportion and appearance of the building.

In order to provide a decorative termination at the top of a masonry wall a course of projecting, moulded stones is often built in. This ornamental projecting stone course is termed a cornice and it is generally laid some one or two courses below the top of the parapet. Fig. 26 is an illustration of a cornice and parapet to an ashlar faced wall.

The parapet wall consists usually of two or three courses of stones capped with coping stones (Vol. 1 Chapter 9) bedded on a D.P.C. of sheet metal. The parapet is usually 1B thick or of such thickness that its height above roof is not more than six times its least thickness. The parapet may be built of solid stone or of ashlar stores bonded to brick backing.

The cornice is constructed of stones of about the same depth as the stones in the wall below, cut so that they project, and moulded for appearance sake. Because the stones project, their top surface is weathered (slopes out) to throw water off.

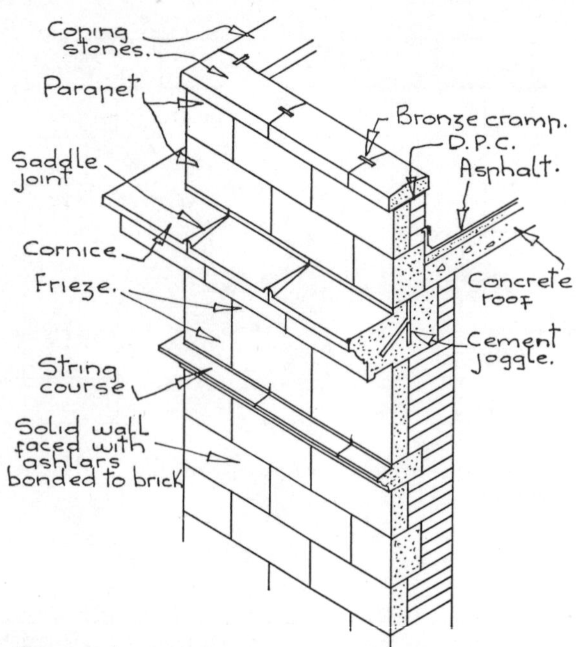


**Fig. 26**

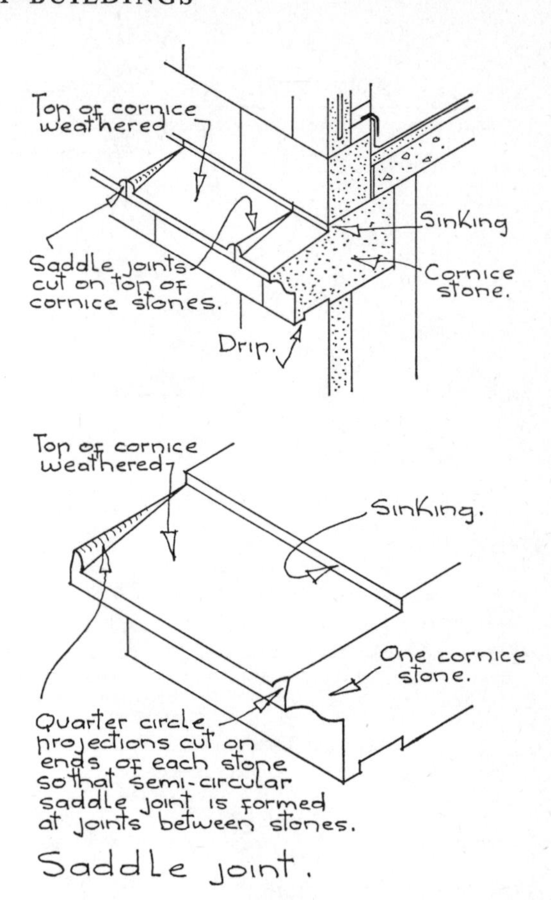


**Fig. 27**

**Saddle joint.** The projecting, weathered top surface of coping stones is exposed and rain running off it will in time saturate the mortar in the vertical joints between the stones. To prevent rain soaking into these joints it is usual to cut the stones to form a saddle joint as illustrated in Fig. 27. The exposed top surface of the stones has to be cut to slope out (weathering) and when this cutting is executed a projecting quarter circle is left on the ends of each stone. When the stones are laid the projections on the ends of adjacent stones form a protruding semi-circular saddle joint which causes rain to run off away from the vertical joints.

**Weathering to cornices.** Because cornices are exposed and are liable to saturation by rain and possible damage by frost it is good practice to cover their exposed top surface with sheet metal. Obviously if the cornice is protected with sheet metal there is no point in having saddle joints formed as described above. Sheet lead is usually preferred as a weathering because of its ductility and impermeability. Fig. 28 illustrates sheet lead weathering to a cornice.

**Cement joggles.** Cornice stones project and one or more stones might in time settle slightly so that the decorative line of the mouldings cut on them would be broken and so ruin the appearance of the cornice. To prevent this possibility shallow "V"-shaped grooves are cut in the ends of each stone so that when the stones are put together these matching "V" grooves form a square hole into which cement grout is run. When the cement hardens it forms a joggle which locks the stones in their correct position. Fig. 29 illustrates a cement joggle.

**Dowels.** To maintain stones in their correct position in a wall slate dowels are used. The stones in a parapet are not kept in place by the weight of wall above as are the stones in the wall below and it is parapet stones that are commonly fixed by means of slate dowels. These dowels consist of square pins of slate which are fitted to holes cut in adjacent stones as illustrated in Fig. 30.

**Cramps.** Coping stones bedded on top of a parapet wall prevent rainwater soaking down into the wall below. But if the mortar in the joints between coping stones cracks, rain will penetrate the cracks to the parapet wall below. If the parapet wall is saturated it is possible that frost may damage it. So that rain can-

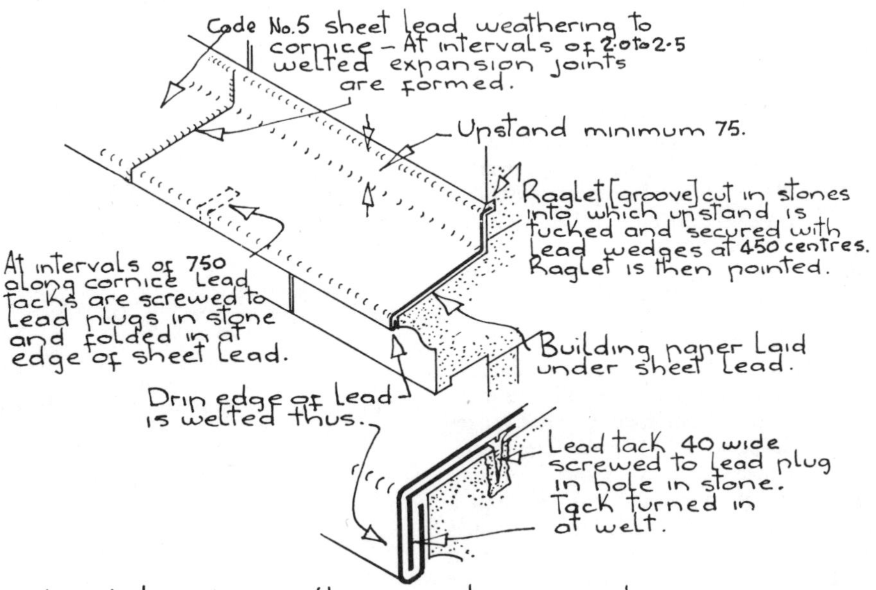
Code No.5 sheet lead weathering to cornice – At intervals of 2·0 to 2·5 welted expansion joints are formed.
Upstand minimum 75.
Raglet [groove] cut in stones into which upstand is tucked and secured with lead wedges at 450 centres. Raglet is then pointed.
At intervals of 750 along cornice lead tacks are screwed to lead plugs in stone and folded in at edge of sheet lead.
Building paper laid under sheet lead.
Drip edge of lead is welted thus.
Lead tack 40 wide screwed to lead plug in hole in stone. Tack turned in at welt.
Sheet lead weathering to projecting cornice.

**Fig. 28**

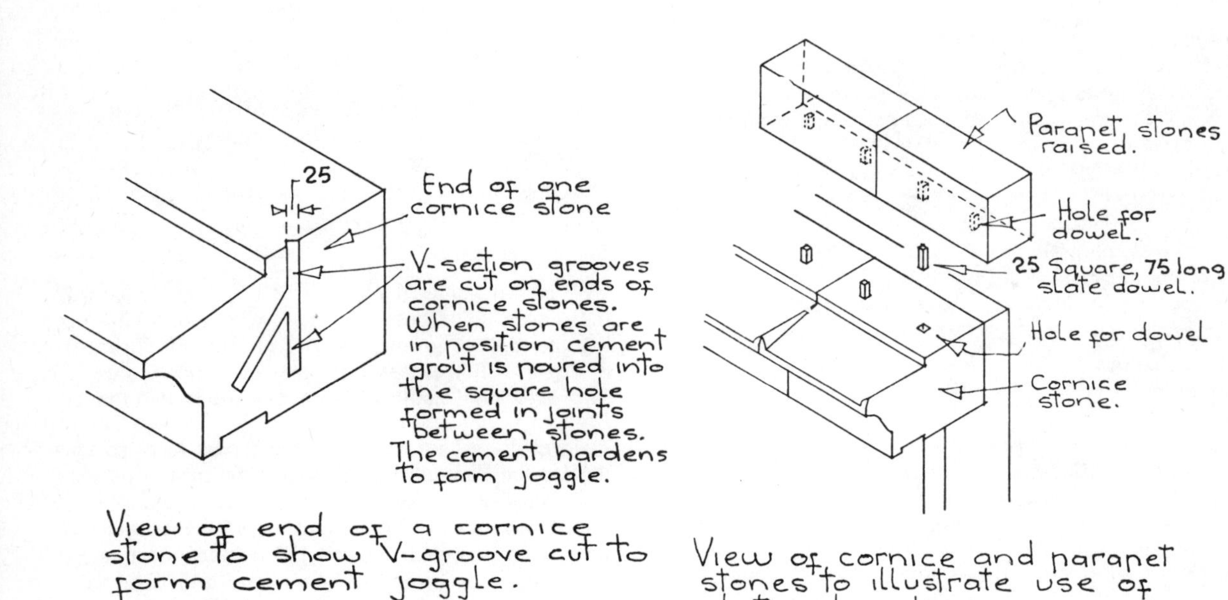
25
End of one cornice stone
V-section grooves are cut on ends of cornice stones. When stones are in position cement grout is poured into the square hole formed in joints between stones. The cement hardens to form joggle.
View of end of a cornice stone to show V-groove cut to form cement joggle.
Parapet stones raised.
Hole for dowel.
25 Square, 75 long slate dowel.
Hole for dowel
Cornice stone.
View of cornice and parapet stones to illustrate use of slate dowels.

**Fig. 29**

**Fig. 30**

not penetrate through cracks between coping stones to the parapet below, it is usual practice today to bed the coping stones on a continuous damp proof course of sheet metal.

The use of a D.P.C. below coping stones is only a recent practice and before a D.P.C. was used masons employed cramps to strengthen the joints between coping stones. These cramps consisted either of a dovetail-shaped piece of slate or a bronze cramp as illustrated in Figs. 31 and 32. If a D.P.C. is built in below coping stones cramps are not absolutely necessary but may be used to maintain the stones in their correct alignment.

**Ashlar masonry joints.** Ashlar stones may be finished with smooth faces and bedded with thin mortar joints, or the stones may have their exposed edges cut to form a channelled or a "V" joint between them. The purpose of channelled and "V" joints is to emphasise the shape of each stone and give the wall a heavier, more permanent appearance than it would have if finished with a plain ashlar face and thin mortar joints. The ashlar stones of the lower floor of large buildings are often finished with channelled or "V" joints and the wall above with plain ashlar stones to give the base of the wall an appearance of strength. Ashlar masonry finished with channelled or "V"-joints is said to be rusticated.

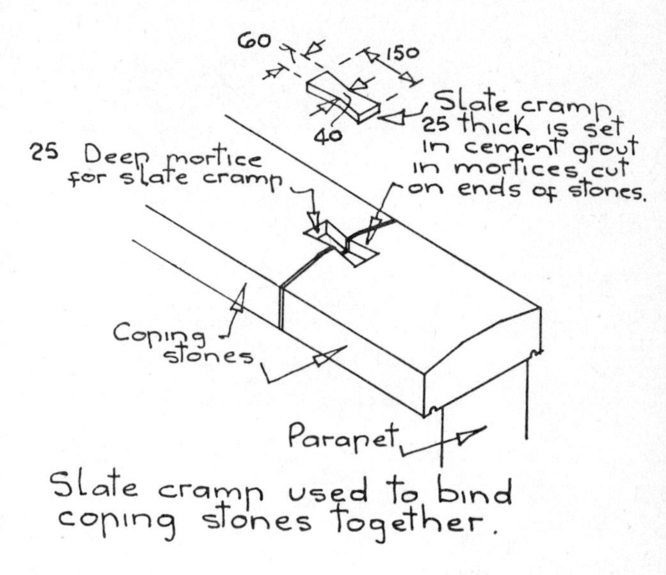


**Fig. 32**

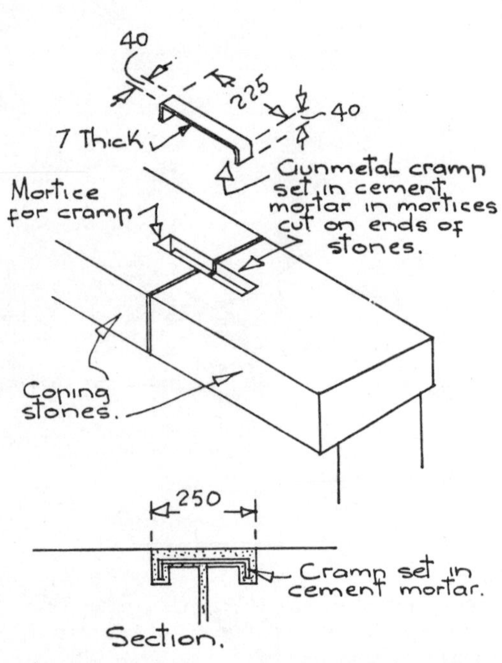


**Fig. 31**

**Channelled joint (rebated joint).** This is formed by cutting a rebate on the top and one side edge of the face of each stone, so that when the stones are put together a channel or rebate appears around each stone as illustrated in Fig. 33.

The rebate is cut on the top edge of each stone so that when the stones are put together rainwater, which may lie in the horizontal channel, will not soak into the mortar joint, as it would if the rebate were cut on the bottom edge of each stone.

**"V" joint (chamfered joint).** This is formed by cutting the edges of stones so that when they are put together a "V" groove appears on face as illustrated in Fig. 34. Sometimes the edges of stones are cut to form a "V" and channelled joint as illustrated in Fig. 35.

Plain ashlar stones are finished with flat faces so that the face of the wall is flat, apart from any projecting mouldings or cornices.
Ashlar stones can also be finished with their exposed face or faces tooled to show the texture of the stone, the other faces being finished square and flat. Some of the tooled finishes used with ashlar stones are illustrated in Fig. 36. It is the harder stones such as granite and hard sandstone that are more commonly finished with rock face, pitched face, reticulated or vermiculated faces. The softer fine grained stones such as limestone are more usually finished as plain ashlar.

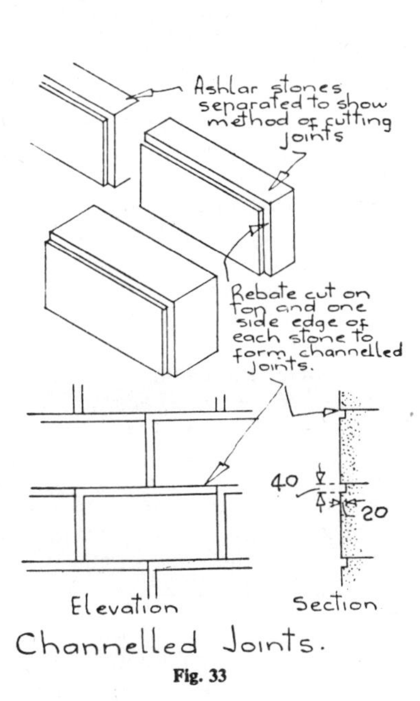


Channelled Joints.

Fig. 33

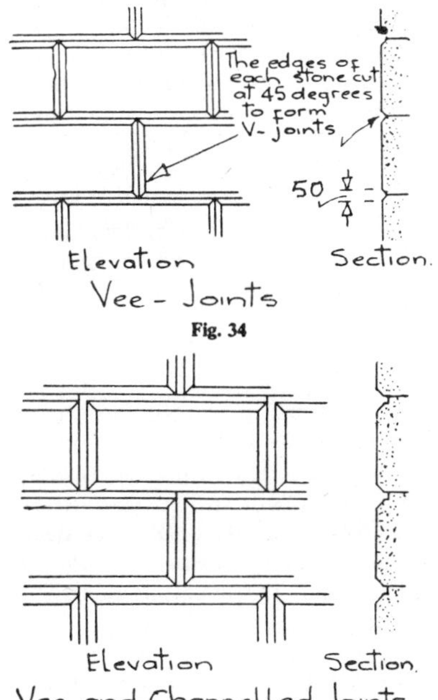


Vee - Joints

Fig. 34

Vee and Channelled Joints.

Fig. 35

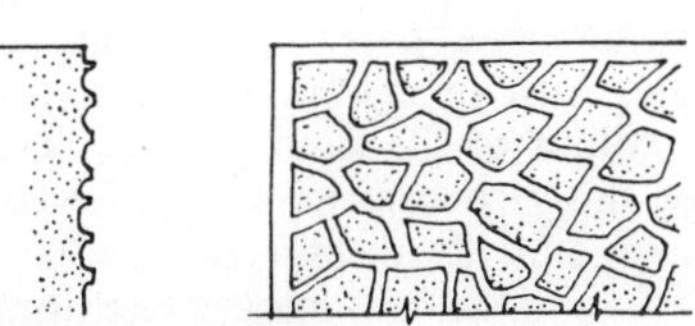

Reticulated surface.
A network of sinkings, 10 deep separated by level bands is cut in the surface.

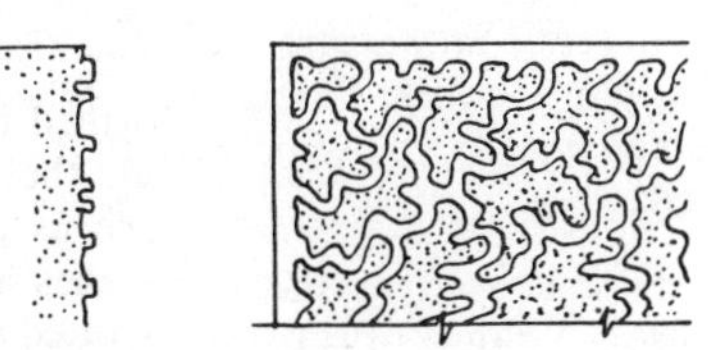

Vermiculated surface.
Worm like sinkings are cut in the face of the stone.

Surface finishes.
Stones with tooled faces are used at quoins, jambs and base of walls. The stones usually have channelled or Vee joints.

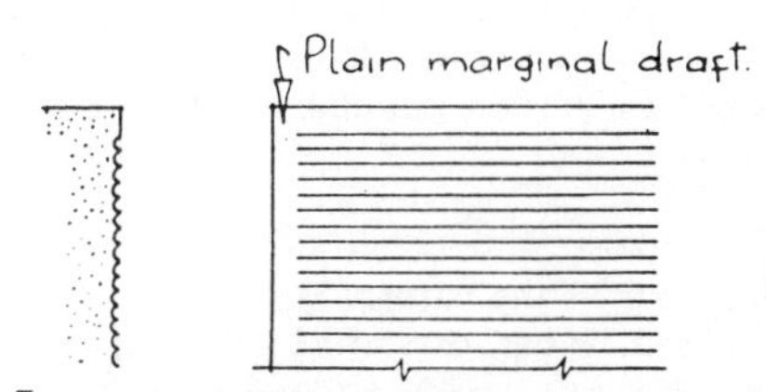


Furrowed surface.
Small flutings 7 or 10 wide cut either vertically or horizontally.

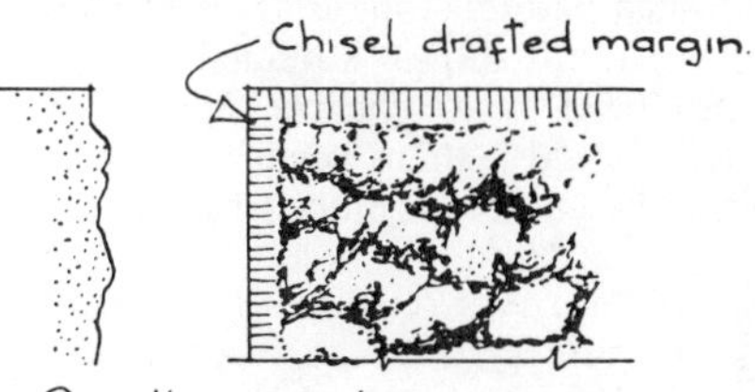


Rock faced surface.
Face of stone worked to resemble natural rock face.

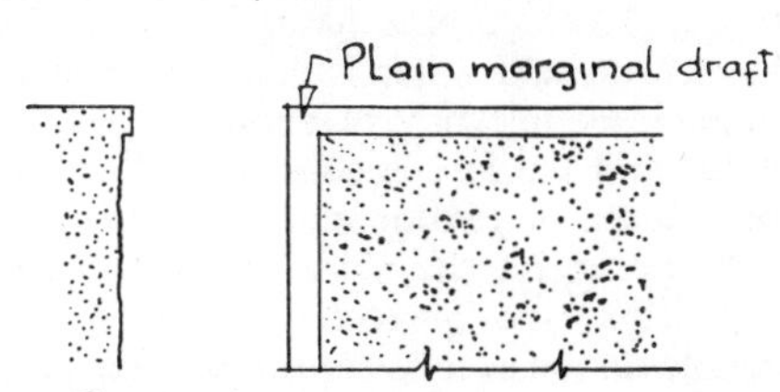


Picked surface.
The face of the stone is picked with sharp point.

Fig. 36

## CAST STONE

In order to get large building stones from a quarry face much broken stone has to be removed. The larger broken stones can be used for rubble walling, but most of the rubble is too small for use in building. Of recent years these small rubble stones have been increasingly used in the manufacture of cast stone. The natural stone is crushed and mixed with cement and water and cast in moulds to form cast stones. At present as much cast stone as natural stone is used in building in this country.

**Reconstructed stone.**

This type of cast stone is made from an aggregate of crushed stone, cement and water. The stone is crushed so that the maximum size of the particles is 15 and it is mixed with cement in the proportions of 1 part cement to 3 or 4 parts of stone. Either Portland cement, white cement or coloured cement may be used to simulate the colour of natural stone as closely as possible.

A comparatively dry mix of cement, aggregate and water is prepared and cast in wooden moulds. The mix is thoroughly consolidated inside the moulds by vibrating and left to harden in the mould for at least 24 hours. The stones are then taken from the moulds and allowed to harden gradually (cured) for twenty-eight days.

Well made reconstructed stone has the same texture and colour as the natural stone from which it is made and it can be cut, carved and dressed just like natural stone. It is not stratified, is free from flaws and is often a better material than the natural stone from which it is made. The cost of a plain stone cast with an aggregate of crushed natural stone is about the same as that of a similar natural stone. Moulded cast stones can often be produced more cheaply by repetitive casting in shaped moulds, than similar natural stones which have to be cut and moulded by hand.

A cheaper form of reconstructed stone is made with a core of ordinary concrete, faced with an aggregate of crushed natural stone and cement. The core is made from clean gravel, sand and Portland cement and the facing from crushed stone and cement to resemble the texture and colour of a natural stone. The crushed stone, cement and water is first spread in the base of the mould to a thickness of about 25, the core concrete is added and the mix is consolidated. If the stone is to be exposed on two or more faces the natural stone mix is spread on the sides and the bottom of the mould. This type of reconstructed stone obviously cannot be carved as it has only a thin surface of natural looking stone. The stones are left to harden in the mould and then cured as previously described.

If reconstructed stone is made from too wet a mix and if the stones are not allowed to harden gradually after being cast, it is likely that hair cracks will appear in their faces. This surface cracking (crazing) is at first unsightly but later, due to the action of frost, the face of the stone may disintegrate. Well made reconstructed stone is as strong and as durable as natural stone.

**Artificial stone**

This type of cast stone is manufactured with a core of concrete faced with sand and cement pigmented to resemble the colour of a natural stone. Artificial stone is cheaper than reconstructed stone because the materials used, gravel and sand, are cheaper than crushed natural stone.

The materials are mixed and cast in wood moulds in the same way as reconstructed stone, and the stones are cured for the same length of time. The facing of pigmented sand and cement is usually 20 to 25 thick.

Because pigmented sand and cement are used for the surface these stones do not usually resemble the colour or texture of natural stone as closely as reconstructed stone.

The surface of artificial stone is liable to crack. Irregular hair cracks appear on the face of the stones due to more pronounced shrinkage in the fine surface material than in the concrete core. The sand and cement facing is often rich in coloured cement to obliterate the natural colour of the sand and the severe drying shrinkage of this rich cement mix causes hair cracks on the surface. Initially the surface cracks are merely unsightly but in time frost may cause the cracks to open up and the facing may then come away from the concrete core.

Cast stone is principally used for ashlar masonry in lieu of natural stone, bonded to brick backing or as a facing to concrete. Stones are cast and used as wall facings, sills, jamb stones, lintels arch stones, cornices and parapets in exactly the same way that the natural stone is used.

Cast stone is also used as the outer skin for cavity walls of houses. A range of standard size reconstructed stones is manufactured from Portland limestone aggregate. The stones are 600 long, ½B wide and 65, 140 or 215 high, with their exposed faces cast to resemble tooled natural stone surfaces. By casting reconstructed stone in standard sizes the cost of a cavity wall constructed with an outer skin of these stones is comparable to that of a cavity wall constructed with facing bricks. These reconstructed stones are sufficiently dense and resistant to weather to be used as the outer skin of a cavity wall in all but exposed positions. Fig. 37 illustrates the use of these stones.

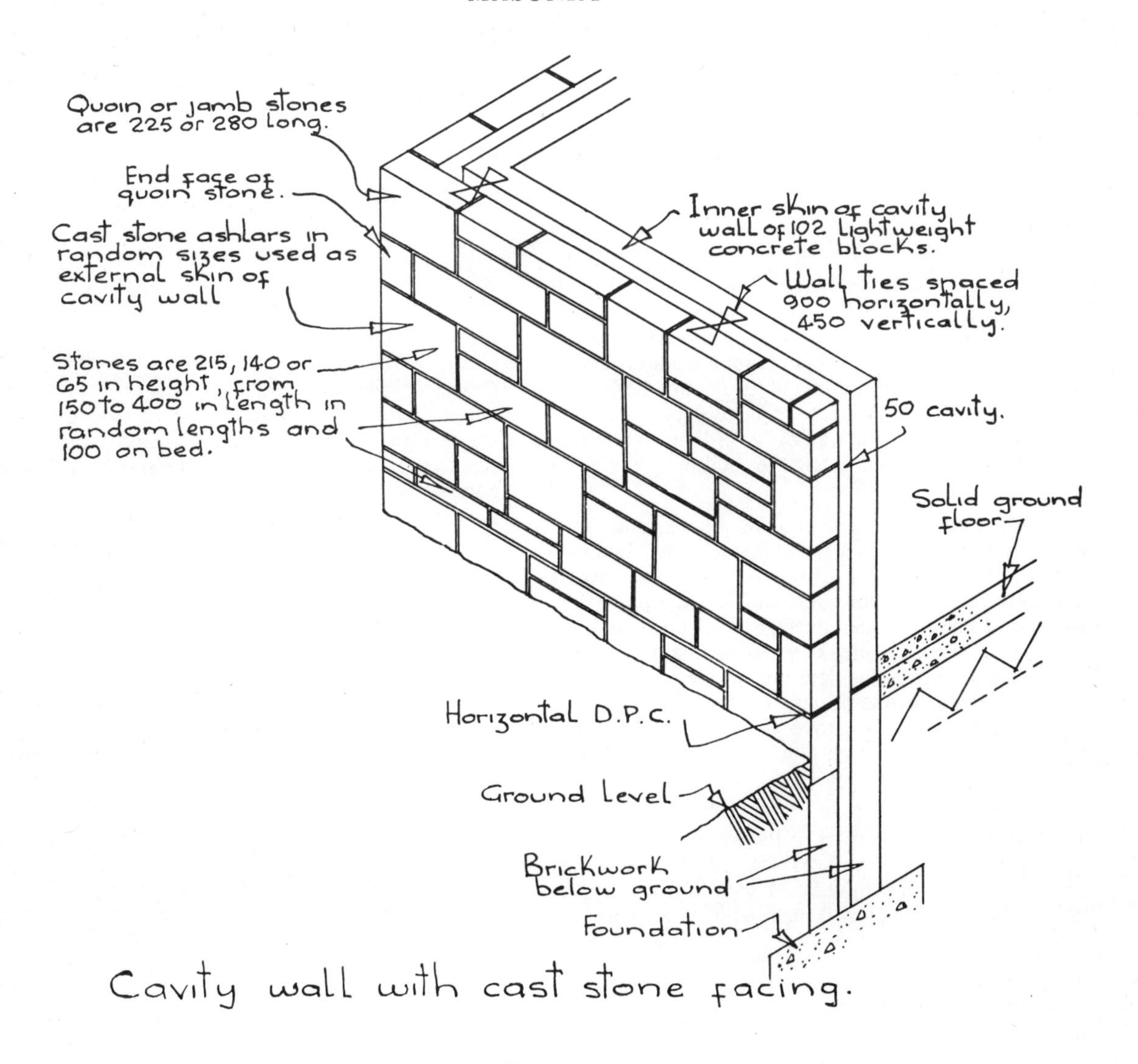


Fig. 37

**Reference Books and Publications**

**British Standards:**
No. 1200. Sands for mortar.
No. 1217. Cast Stone.
No. 1240. Natural stone lintels.
No. 3798. Coping units.
No. 4374. Sills of clayware, cast concrete, cast stone, slate and natural stone.

**British Standard Code of Practice:**
CP.121.201. Masonry walls ashlared with natural stone.
CP. 121.202. Masonry-Rubble walls.
**Building Research Station Digests:**
Nos. 20 & 21. The weathering, preservation and maintenance of stone masonry. (First Series).
Building Stones by John Watson. Published by Cambridge University Press.

## CHAPTER TWO

# WINDOWS

## SfB (31) Windows: General

Windows are disigned to allow adequate natural light into buildings. The Building Regulations 1965 stipulate minimum sizes of windows for lighting habitable rooms. (G.L.C. Constructional By-laws require windows equal to at least one tenth of the area of the floor of the room). Obviously the size of window required depends on the use of the room that the window gives light to and the position of obstructions facing the window.

Windows also serve to ventilate rooms and The Building Regulations 1965 and the Greater London Council Constructional By-laws state that every habitable room shall be provided with a window or windows which shall open directly to the open air, and constructed so that an area of window equal to not less than one twentieth of the floor area of the room can be opened. The top of the opening part of windows must be not less than 1·75 above the floor. (In G.L.C. By-laws this figure is given as 1·981 in top storey and 2·134 in every other storey).

The manner in which a part or the whole of a window is arranged to open affects its construction and appearance. The three most usual ways in which a part or the whole of a window is made to open are with

(a) side hinged casements,
(b) pivoted sashes, and
(c) vertically sliding sashes.
(d) horizontally sliding sashes.

Fig. 38 illustrates the three different types of window. The following is a description of the general arrangement and advantages of each.

**Casement window:** The simplest form of casement window consists of a square or rectangular window frame of wood or metal, with a casement hinged at one side to the frame to open out. The side-hinged opening part of the window is termed the casement and it consists of glass surrounded and supported by wood or metal. Fig. 39 is a view of a simple casement window.

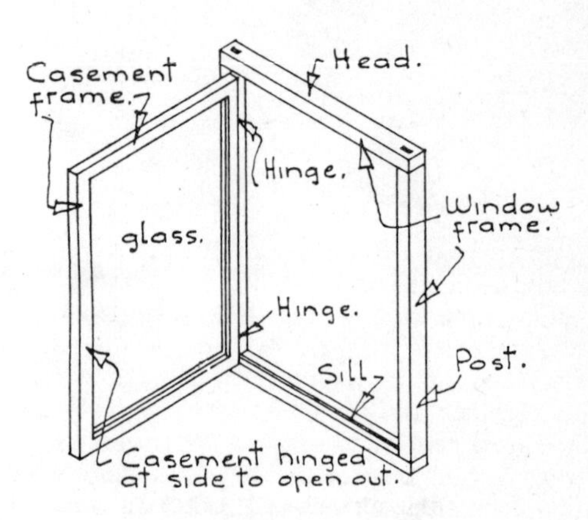


**Fig. 39**

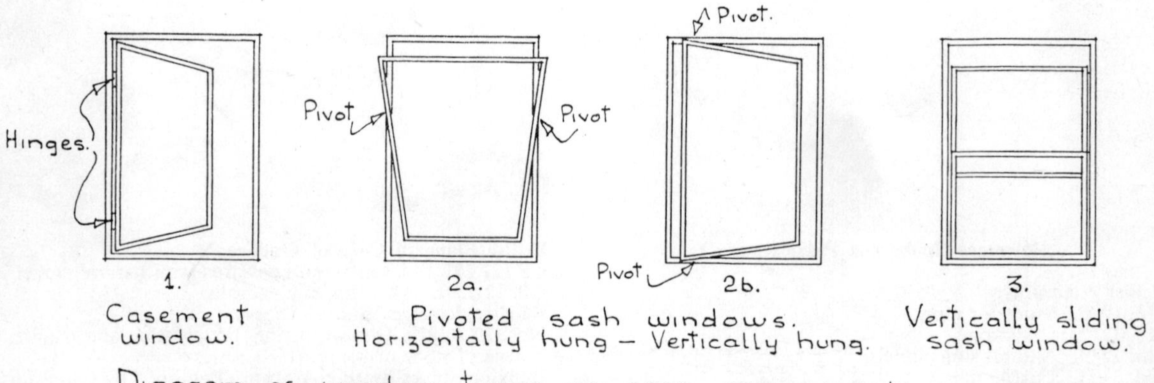


**Fig. 38**

The casement is hinged to open out because an outward opening casement can more readily be made to exclude rain and wind than one opening inwards. The reason for this will be explained later. Because a casement is hinged on one side, its other side tends to sink, due to the weight of the casement when open. This is illustrated in Fig. 40. If any appreciable sinking occurs the casement will bind in the window frame and in time may be impossible to open. Obviously the wider a casement the greater its weight and the more likely it is to sink as described. It is generally considered unwise to construct casements wider than say 600 If a casement window is wider than 600 it will consist of two or more casements.

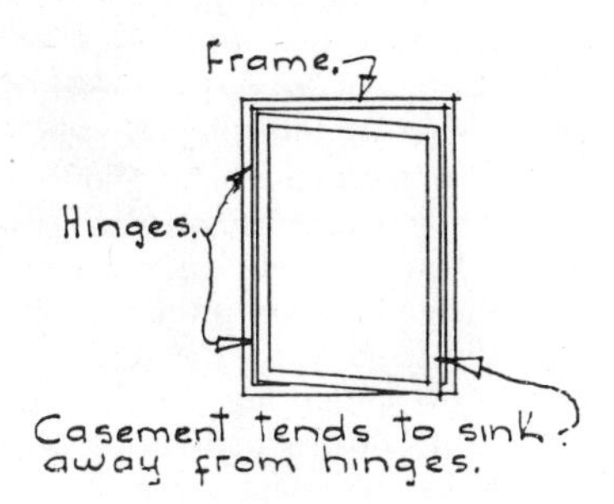


Fig. 40

A window with two casements can be designed with the casements hinged so that when closed they meet in the middle of the window. This is illustrated in Fig. 41. The disadvantage of this arrangement is that due to expansion or sinking, or both, the casements may in time jam together and be difficult to open. It is usually considered better to construct the window frame with vertical wood or steel members, called mullions, to which each casement closes. This is illustrated in Fig. 42. In winter it is sometimes difficult to open a casement

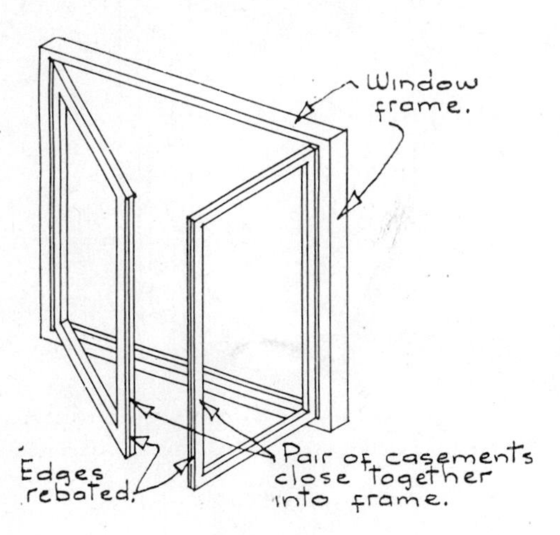


Fig. 41

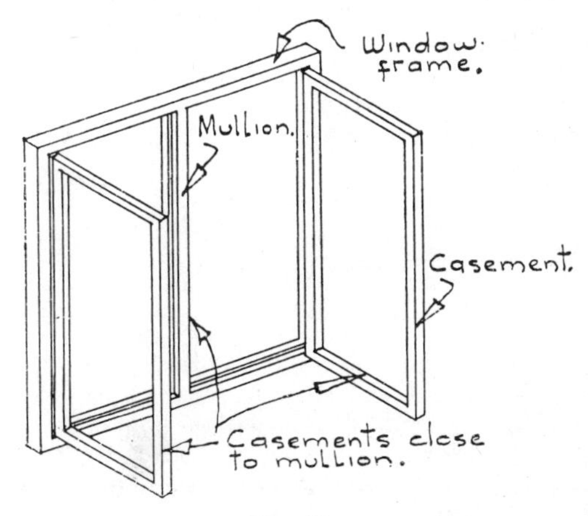


Fig. 42

just enough to ventilate a room without letting in too much cold air, so most casement windows are designed with small opening parts above the casements. These are called ventlights and are usually hinged at the top to open out. So that the ventlights can be opened independently of the casements the window frame is constructed with a horizontal member, called a transom, to which ventlights and casements close. Fig. 43 illustrates a window with casements and ventlights. Casement windows with ventlights are usually designed so that the transom is above the average eye level of people using the room, for obvious reasons.

Windows which have two or more casements are sometimes described as being two, three or four-light

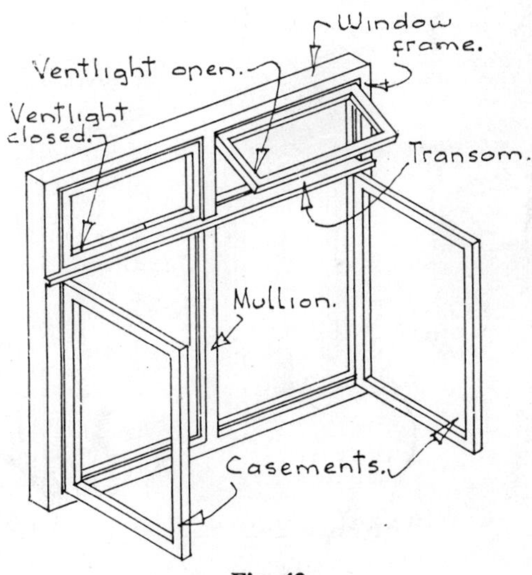


Fig. 43

casements as illustrated in Fig. 44. The word light referring only to the number of casements. By itself the description four-light casement window is too vague to be of much use and it is better to describe windows by the number of casements and ventlights in them.

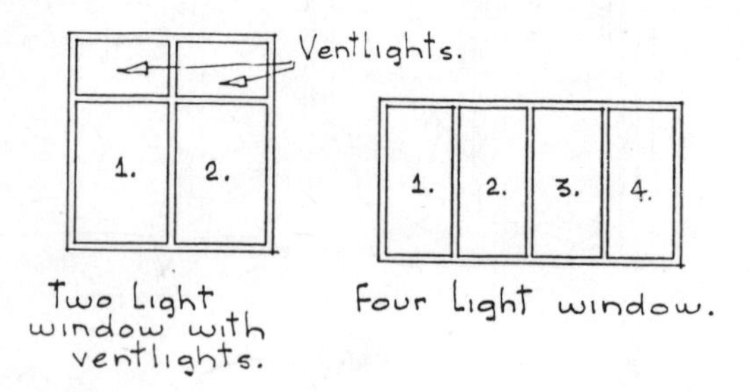


Fig. 44

**Dead lights:** Many casement windows are constructed so that only a part of them can be opened. For example one standard type of metal and wood casement window is made with one casement and one ventlight, the rest of the window being of glass fixed in the window frame, as illustrated in Fig. 45. The part of the window which cannot be opened is called a dead light.

Most casement windows used in new buildings today are of standard steel or wood sections made up in various standard arrangements of casements, ventlights and dead lights.

Casement windows are cheaper than other types of window of similar size. Their moving parts, hinges, are simple and unlikely to fail and these windows are more commonly chosen than other types.

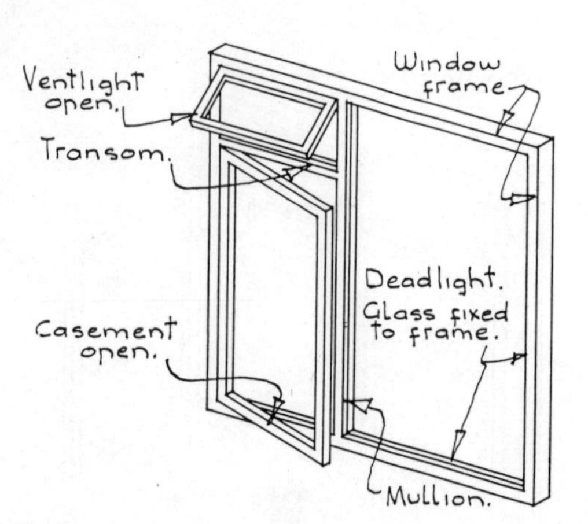


Fig. 45

The disadvantage of a casement window is that the casements and ventlights, and mullions and transoms reduce the possible unobstructed area of glass and therefore light in a window of any size. Of recent years it has been fashionable to use windows with as large an unobstructed area of glass as possible and the casement window with its mullions and transoms and comparatively small casements has lost some favour. The manufacturers of standard casement windows now make a range of windows which combine a large dead light with a casement alongside it and a ventlight above.

**Pivoted sash windows:** The opening part or parts of these windows are supported by pivots each side or at the top and bottom of the window frame, so that they open partly into and partly out of the room as illustrated in Figs. 46 and 47. The word sash is used to describe the opening part of these windows and the sash includes the glass and its wood or metal surround. If the sash is pivoted at the sides the pivots are usually fixed slightly above the centre of the height of the window so that the sash tends to be self-closing. Pivots at the top and bottom of sashes are usually fixed away from the centre of the width of the window so that more of the sash opens out than in, so as not to obstruct space inside rooms. The window frame and sash or sashes can be constructed of wood or metal.

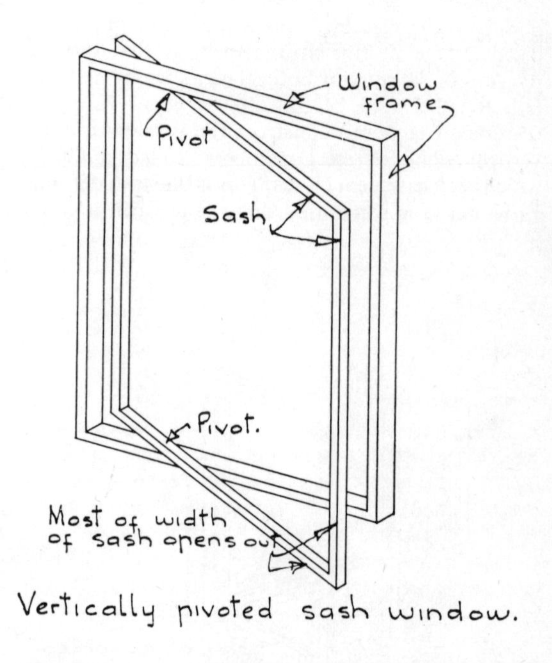


Fig. 46

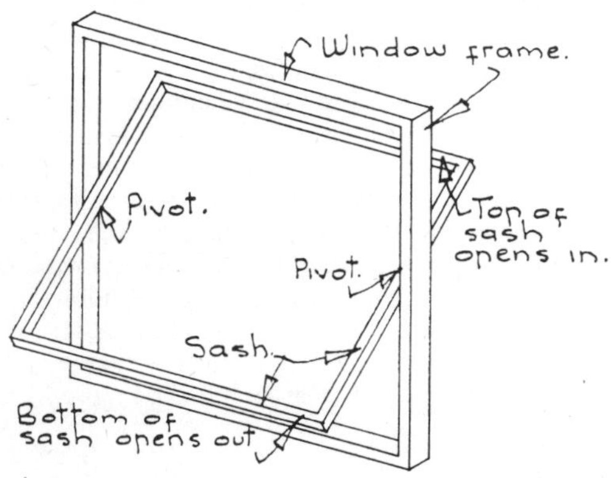


**Fig. 47**

Because the weight of a pivoted sash when open is borne equally by each side pivot, it is less likely to lose shape than a casement, and sashes up to say 1·5 or 1·8 wide can safely be constructed. Because of this a pivoted sash window can consist of one large square of glass. This is an attractive feature of this type of window as it at once lets in more light and offers less obstruction to the view of those looking out than does a casement window of similar size. Another advantage of pivot-hung sash windows is that they can more easily be fitted with double glazing than either casement or sliding sash windows. Double glazing describes the fitting of two sheets of glass spaced a short distance apart in each sash to reduce heat loss through windows. Double glazing will be explained later in detail.

Another advantage of pivot-hung sash windows is that the whole of the area of the window can be opened, whereas sliding sash windows can only be partially opened. A further advantage of these windows is that they can be made so that the sash can be turned through 180 degrees and it is possible to clean the glass on both sides from inside the building. The one disadvantage of this type of window is that it is sometimes difficult to open just sufficiently to allow ventilation without causing draughts. This difficulty can be overcome by forming small ventlights above or below the pivoted sash. These will be top or bottom-hinged, to open out and in respectively. As with casement windows if there are two or more pivoted sashes in a window they will close into vertical mullions which are a part of the window frame, and if there are ventlights, a horizontal transom, which is part of the window frame, will be used.

At present pivoted windows are generally more expensive than standard casement windows of similar size, mainly because the standard casement window is mass produced on a large scale which makes for cheapness. No doubt the many advantages of the pivoted sash window will become more generally known in time, and this type of window will be mass produced as cheaply as casement windows.

**Sliding-sash windows**

**Vertically sliding-sash window (Double hung sash):** The word sash is used to describe an opening part in this type of window. The term sash describes a wood or metal frame which contains and supports the glass together with the glass itself, and any glazing bars there may be. The most commonly used type of sliding-sash window is that in which the sashes are hung to slide vertically inside the window frame. This type of window is associated with buildings erected in England during the period 1700 to 1900, and is an integral part of the Architecture of that period.

Until recently these windows consisted of two sashes, usually of equal size, hung inside a box-like timber window frame, termed a cased frame. The two sides of the frame were made in the form of a box (cased) so that metal weights could be suspended inside them. Each sash was hung on two cords which ran over metal pulleys in the side of the frame and were attached to the weights inside the cased frame. If the two weights combined were as heavy as the sash little effort was required to raise or lower the sash. Fig. 48 illustrates the general arrangement of this type of window.

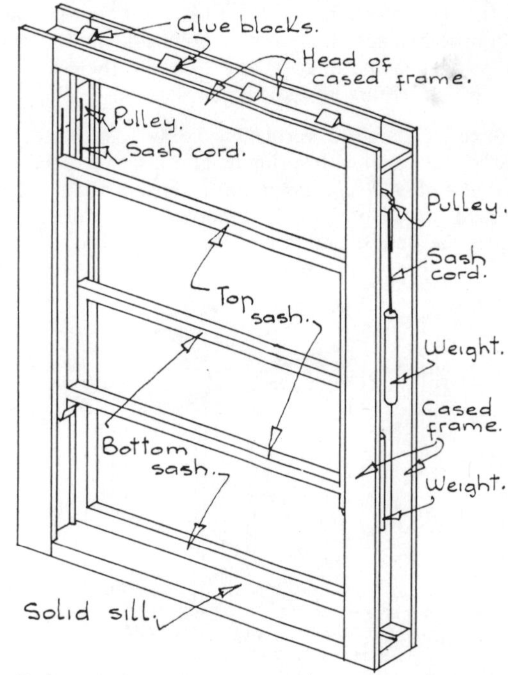


**Fig. 48**

The principal advantage of this type of window is that as the sashes are suspended vertically they do not tend to sink and lose shape as casements do, and in consequence sashes as wide as 1·5 can safely be constructed with very small sections of wood or metal. Apart from the central horizontal sash members, where the sashes meet when closed, the window can be almost entirely of glass inside the window frame. The appearance of this type of window is naturally elegant and it is associated with one of the periods of more attractive domestic building in this country.

Another advantage of this type of window is that it can be opened in cold weather just sufficiently to ventilate rooms without letting in draughts of cold air.

One of the disadvantages of the sliding-sash window is that unless it is well maintained the sashes are liable to jam inside the frame. Obviously the counterbalance weights each side of a sash should be of equal weight, otherwise one side of the sash will rise more readily than the other and the sash will jam inside the frame. In time sash cords become frayed and need renewing and it is generally beyond the capabilities of an average householder to do this.

Of recent years metal spring balances have been on the market for use with this type of window instead of cords and weights. These balances are fixed either to each side of the sashes or to the frame. Their advantage is firstly that the window frame need not be cased, as there are no counterbalance weights to run inside it and this effects some saving in cost, and secondly there are no cords which require periodic renewal.

A range of standard vertically sliding sash windows in wood and metal with spring balanced sashes is manufactured today. They are more expensive than either casement or pivoted sash windows of similar size because their construction is more complicated.

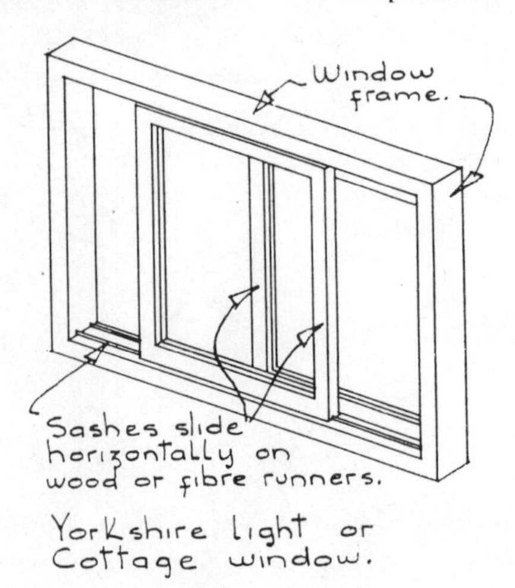


Fig. 49

**Horizontal sliding-sash windows:** These consist of a window frame of wood or metal in which are at least two sashes, one or both of which can be opened by sliding horizontally. Fig. 49 illustrates the general arrangement of this type of window. The sashes are made to slide on wood, metal or compressed fibre runners fixed inside the frame. This type of window is sometimes known as a Yorkshire light because for many years cottages in that part of the country were built with this type of window. This type of window is very little used today because it suffers from the disadvantage that unless it is very accurately made the opening sashes are likely to jam. The larger the sash the more likely it is to jam in the frame.

**Glazing bars:** Glass for windows was originally made by hand and it was then impossible to make large sheets of glass. A casement or sash had to be divided by a number of glazing bars into which the small panes of glass could be fixed. Today glass is manufactured in large sheets and there is no longer any necessity for glazing bars. The fashion now is for casements and sashes without glazing bars.

To describe the construction of the three types of window in detail.

**Casement window**

**Standard steel casements:** A window commonly used today is the standard steel casement. It is as cheap as other types of window and is made in a wide range of standard sizes. These windows are made from a standard section of mild steel, which is used for the frame, casements and ventlights, and is illustrated in Fig. 50.

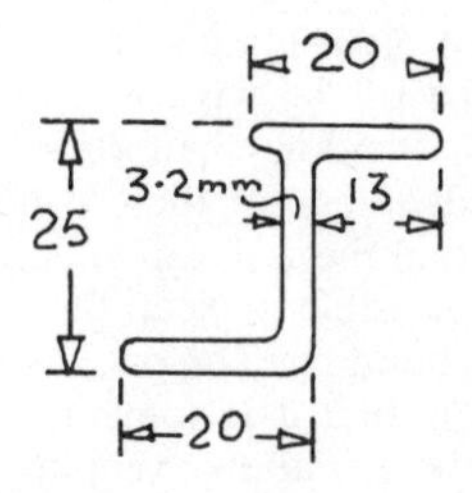


Fig. 50

It will be seen that the section is roughly "Z" shaped. Lengths of this section of mild steel are cut, trimmed and welded together to form a range of standard sized windows. The section of steel from which these windows are made is designed so that the casements fit conveniently in the frame as illustrated in Fig. 51.



write the sizes out of the notebook.

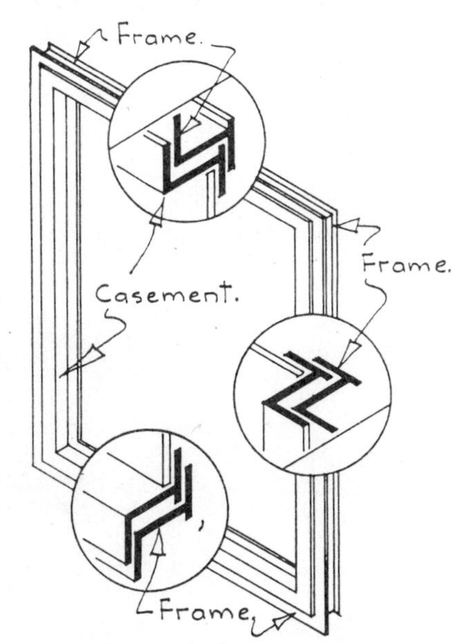


Standard metal casement showing enlarged details at head, jamb and sill.

**Fig. 51**

Where there are two casements, or a casement with a dead light, a mullion is welded into the frame. With ventlights in the window a transom is welded into the frame. This is illustrated in Fig. 52 which is a view of a standard metal casement window.

By using the standard section for making frames, casements and ventlights, and the standard mullion and transom sections, a wide range of standard window sizes can be made up.

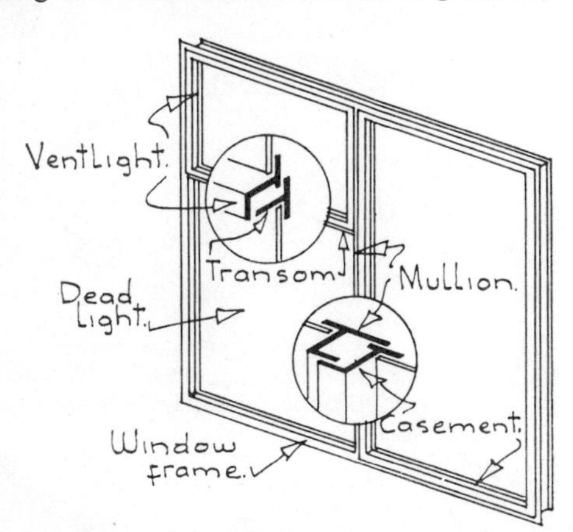


Standard metal window with casement, ventlight and deadlight.
Showing enlarged details at mullion and transom.

**Fig. 52**

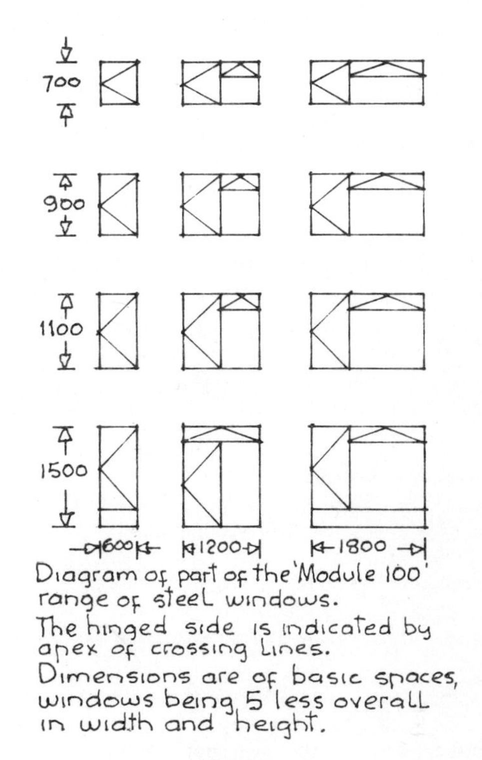


Diagram of part of the 'Module 100' range of steel windows.
The hinged side is indicated by apex of crossing lines.
Dimensions are of basic spaces, windows being 5 less overall in width and height.

**Fig. 53**

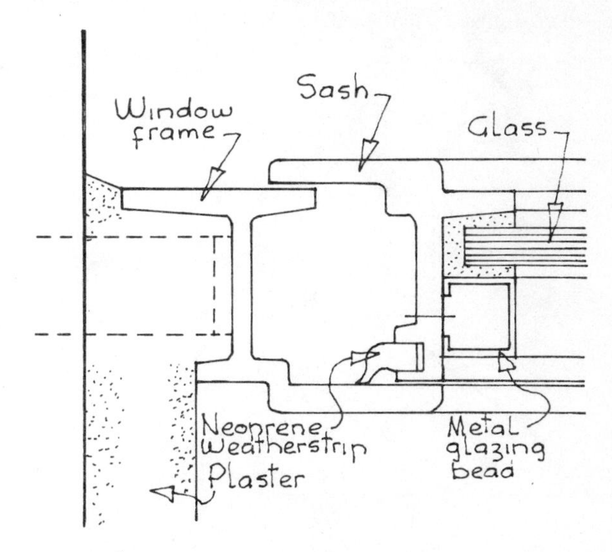


Metric W.20 steel frame and sash window sections.

**Fig. 54**

For the change over to metric the Steel Window Association have proposed to the British Standards Institution two ranges of dimensionally co-ordinated steel windows based on the 100 module. The 'Module 100' range is designed primarily for domestic buildings and the 'Metric W20', which will be made from heavier sections, for buildings whose cost does not have to be kept to a minimum.

Fig. 53 is an illustration of a part of the 'Module 100' range of standard steel windows which will be available ex stock in January 1971, They will be manufactured from the section illustrated in Fig. 50 with the mullion and transom sections illustrated in Fig. 52.

Fig. 54 illustrates typical 'Metric W20' steel frame and sash sections. A neoprene strip is fitted to the sash to reduce air infiltration and heat loss. These windows are designed for glazing from inside, for easy glass replacement, the glass being secured with metal beads.

**Hinges and fasteners:** Standard metal casements are hung on projecting steel hinges as illustrated in Fig. 55. The reason for using projecting hinges is to make it possible to clean the glass in the casement on both sides from within the building. When the casement is opened it is possible for a cleaner to get an arm between the frame and casement and so clean the outside of the glass.

Casements are secured with a single pivoted cock spur type fastener and a peg stay is fixed to the frame and casement so that the casement can be kept open in windy weather, as illustrated in Fig. 56. Ventlights are hung on ordinary steel hinges and are fitted with a peg stay similar to that used for casements.

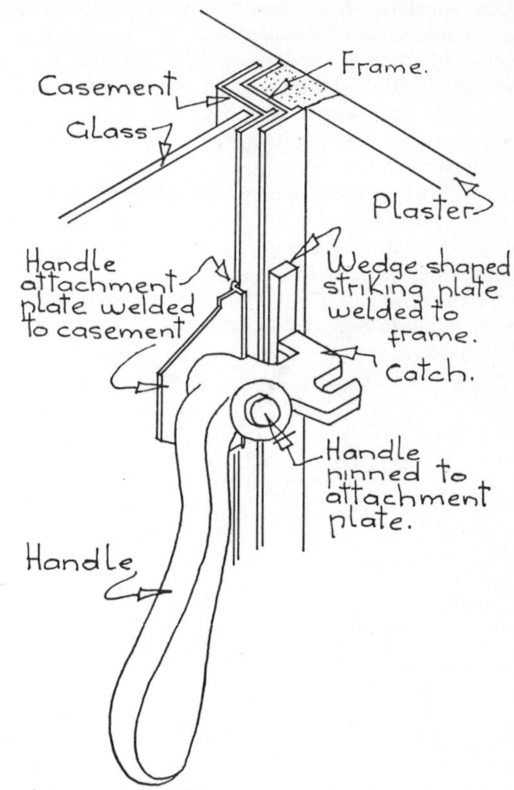


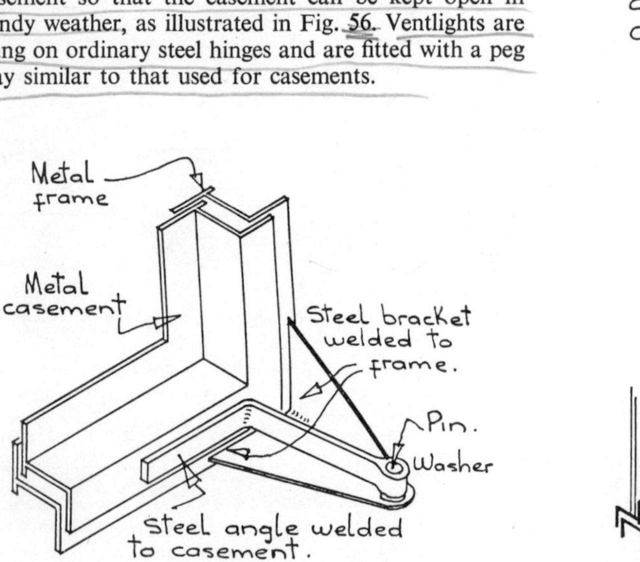

Fig. 55

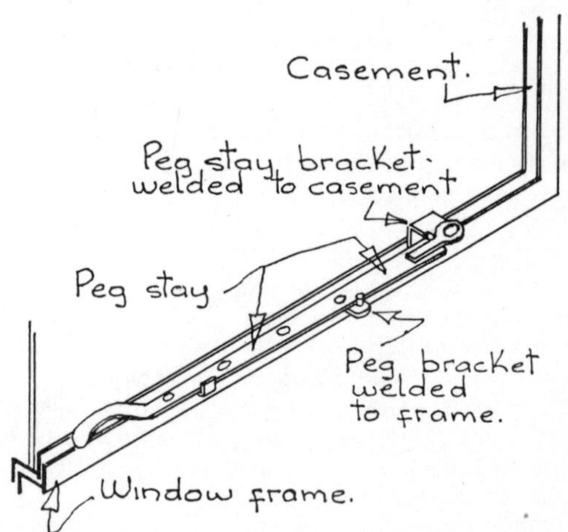


Fig. 56

**Fixing standard steel windows:** Standard steel casement windows are usually "built in" to openings in brick or block walls. The expression "built in" denotes that the window is placed in position when brick or block walls have been raised to sill level and the brickwork or blocks are built around the window frame. The alternative to building in windows is to fix them after the brick or block wall has been built. The purpose of doing this would be to avoid damage to expensive bronze or hardwood windows which might occur while brickwork is being built around them.

Standard steel casement windows are supplied with steel building-in lugs and countersunk headed bolts and nuts. The lugs are "L" shaped strips of galvanised steel as illustrated in Fig. 57.

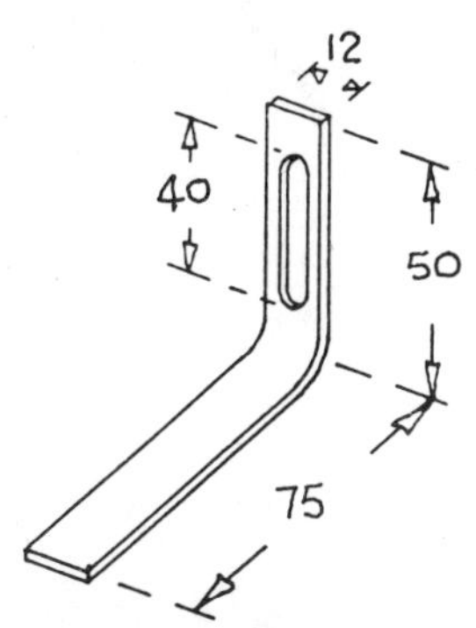


View of galvanised steel lug for building standard metal windows into brickwork.

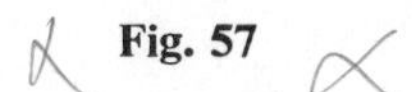

Fig. 57

A set number of building-in lugs is supplied for each particular size of window and they are bolted to the back of the window frame through holes ready drilled in the frame. The lugs are adjustable so that before they are finally tightened to the frame they can be slid up or down the back of the frame so that the long arm of the "L" coincides with a horizontal brick course as illustrated in Fig. 58. Small windows are supplied with building-in lugs for fixing to the side of frames only and large windows with lugs for fixing to all four sides of the window frame.

The building-in lugs supplied with standard steel windows are too small to secure the window firmly in brick or block walls and as an additional fixing the makers recommend that a cement and sand fillet be run between the back of the frame and brick or block jambs as illustrated in Fig. 59.

The fillet is made with a mixture of one part of cement to three of clean sand and is run all round the frame. It effectively secures the window frame in position. There would be much sense in the manufacturers of these windows supplying more substantial building-in lugs than they do.

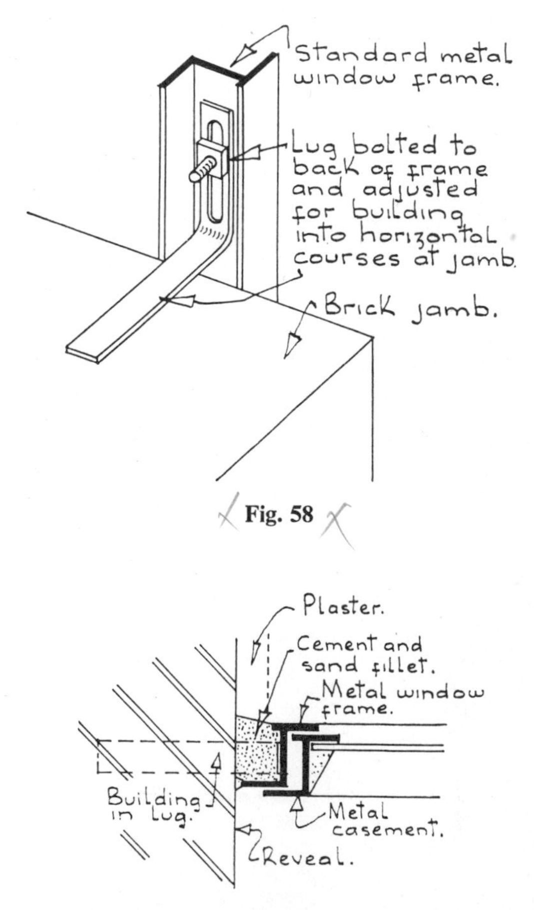


Fig. 58

Fig. 59

The cheapest of the standard metal windows is made from mild steel. Mild steel is used because it can readily be rolled into the section required and because of its good compressive and tensile strength. But mild steel rusts on exposure to air and the metal windows which were first made caused a great deal of trouble, as even with a good paint film they soon rusted and became unserviceable.

**Rustproofing steel windows:** The majority of steel windows today are delivered with a rustproof zinc coating. Zinc is used as a rustproof coating to steel because it corrodes more slowly than steel and also because even if damaged slightly it protects the steel sacrificially. If the zinc coating is damaged an electrolytic action may be set up which causes corrosion of the zinc but prevents corrosion of the exposed steel. As the zinc corrodes to the benefit of the steel near it, it is described as a sacrificial coating.

The methods of coating mild steel windows with zinc are:

**(a) Hot dip galvanising:** This is the most commonly used method. It consists in dipping the cleaned window frame, casement and ventlights, before they are put together, in a bath of molten zinc which adheres strongly to them in the form of a thin coating.

**(b) Electro-galvanising:** The cleaned parts of the window are electroplated in a solution of zinc salts. The window part acts as cathode on which a coating of zinc forms from the zinc anode.

**(c) Sherardizing:** The cleaned parts are placed in a container with zinc powder. The container is sealed and rotated in a furnace whose temperature is raised to 375°C. After some hours at this temperature the container is allowed to cool slowly before removing the contents, now coated with a uniform adherent film of zinc.

**(d) Zinc spraying:** A fine spray of molten zinc is blown on to all the surfaces of the window parts to form a zinc coating.

Because the colour and texture of a zinc coating is not attractive these windows are usually painted. As well as being decorative the paint film protects and generally increases the life of the zinc coating below it. Most paints do not adhere well to hot dip galvanised surfaces and many manufacturers phosphate the galvanised surface so that paint will adhere to it.

Sherardized and zinc sprayed coatings form a good surface for paint and need no treatment.

Electro-galvanised surfaces generally take paint quite well if they have been exposed to the atmosphere for some time.

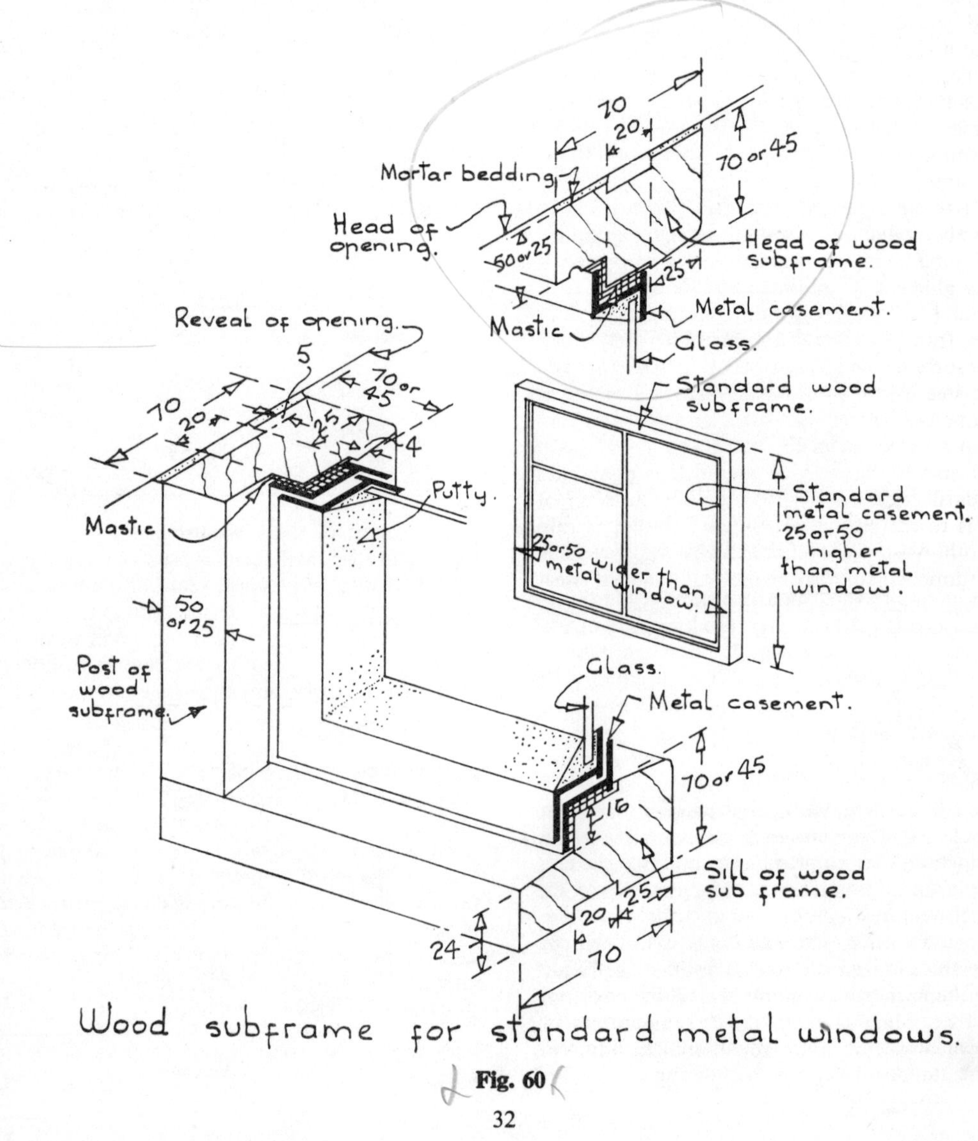


Wood subframe for standard metal windows.

Fig. 60

**Position of window frame in width of reveal**

The standard steel window frame can be fixed at the external face or internal face of a wall or in any position between the two in the width of the reveal of the opening. The position in which it is fixed is partly a matter of taste and partly determined by the construction of the sill of the window. The description of sill construction given later on will explain and illustrate this.

**Timber sub frame to metal windows**

The standard section of the 'Module 100' range of windows is somewhat light and to prevent the window frame or casements being twisted either in transit, during handling on the site or due to binding of casements in the window frame a timber sub-frame may be used. The steel window manufacturers will supply these wood surrounds with their standard steel windows and it is an admission on their part of want of strength in their steel windows that they do so. Another reason for using a wood sub frame is that it improves the appearance of a steel window by providing a more substantial frame to the window which otherwise looks rather weak and inconspicuous.

The wood surrounds are usually cut from 75 x 75 or 75 x 50 softwood timber which is wrought (planed smooth), rebated and joined with mortice and tenon or dowelled joints solidly glued. Fig. 60 illustrates the section of a wood surround with steel window in position. The steel window frame is secured to the timber sub frame with countersunk headed wood screws driven through holes in the steel frame and mastic is packed between the steel frame and the timber sub frame to keep rain out. The two rebates cut in the sub frame are of such depth and so spaced that they accommodate the flanges of the standard steel window frame. The purpose of the rebates is to act as a check against wind and rain that might otherwise seep behind the steel frame. The wood sub frame is supplied with "L" shaped steel building-in lugs which are screwed to the back of it and are built into horizontal joints in the brickwork.

**Steel sub frames for steel windows**

Like timber sub frames, steel sub frames are used to strengthen steel windows and also to provide a wider frame to the window for appearance sake. Two profiles of steel sub frames are manufactured for use with standard steel windows, one for use in solid walls and the other for cavity walls. The frames are manufactured from 1·6mm thick mild steel strip pressed or rolled to the required shape. The head, jamb and sill members are welded together. The steel sub frame is coated with a rustproof zinc coating after manufacture. Fig. 61 illustrates the standard sections manufactured.

The sub frames are built in and secured with adjustable building-in lugs bedded in the horizontal joints of solid brickwork (Fig. 62) or secured by means of the back flanges built into the cavity of cavity walling, as illustrated in Fig. 63. It will be seen that the sub frame closes the cavity at jambs in cavity walling and a D.P.C. is not necessary. The arrangement at the head and sill of the window is illustrated in Fig. 63.

Cavity wall type — Solid wall type.

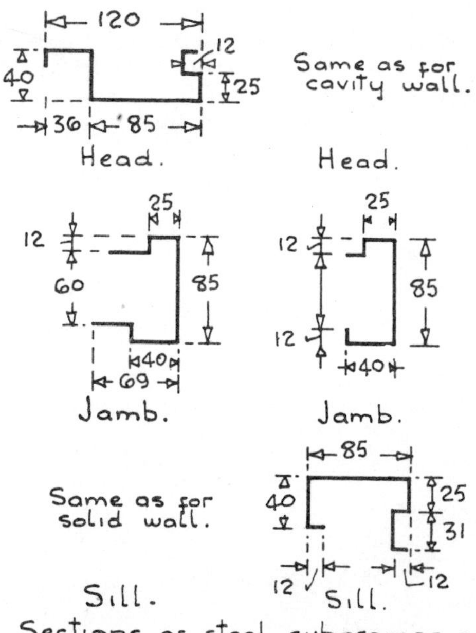


**Fig. 61**

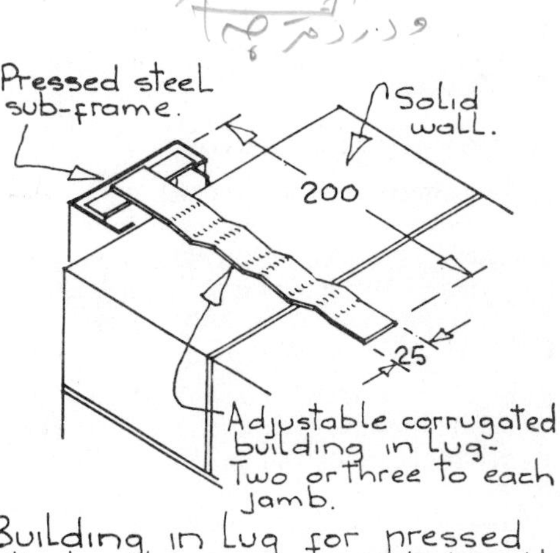


Building in lug for pressed steel sub frame in solid wall.

**Fig. 62**

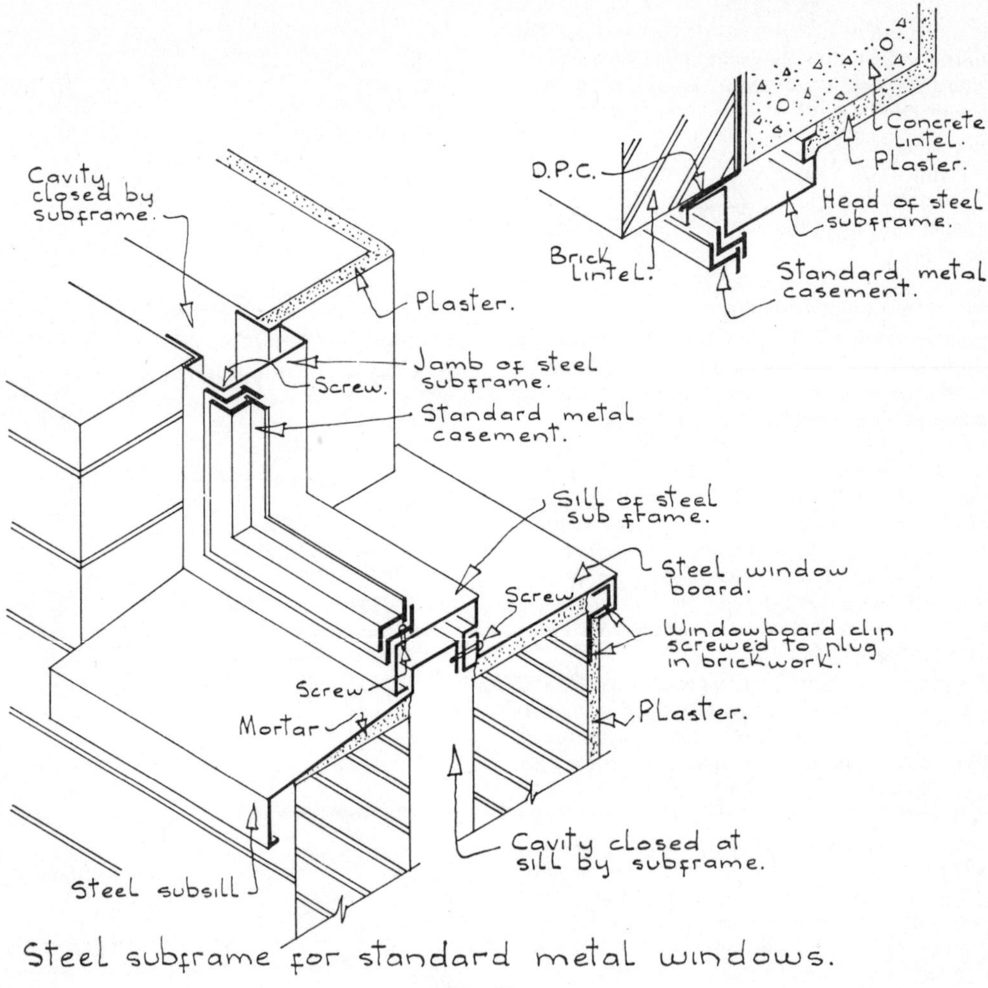


Fig. 63

The standard steel window is secured to the steel sub frame with self tapping screws driven through the window frame into holes ready drilled in the sub frame.

### Steel window sills and window boards

A range of standard steel window sills is manufactured for use with standard steel windows and these are used in lieu of clay, stone or brick sills externally, and tile or wood internally. The standard sections produced are illustrated in Fig. 64. These sills can be used for standard steel windows with or without sub frames.

In Fig. 63 a steel window board is shown to finish the internal sill. The section of steel sub sill used determines the position of the window in the depth of the reveal as illustrated.

### Wood casement windows

For centuries wood casement windows have been used in buildings. Before the twentieth century glass was

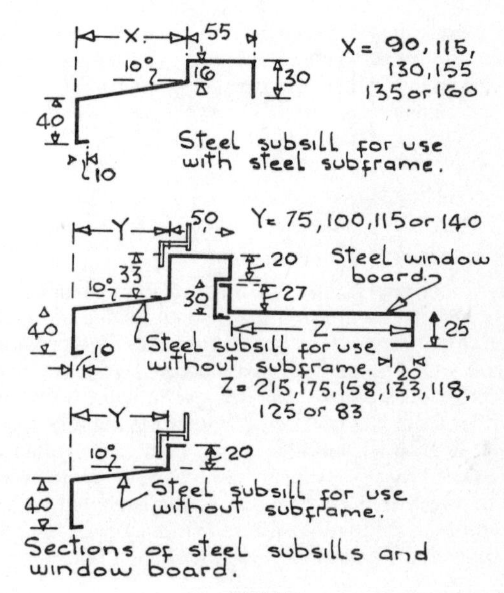


Fig. 64

expensive and could only be manufactured and cut into small squares or diamond shapes and small casements with glazing bars were adequate to the limited area of glazing practicable.

A casement window frame consists of a head, two posts (or jambs) and a sill joined with mortice and tenon joints, together with one or more mullions and a transom, depending upon the number of casements and ventlights.

The members of a wood window frame are cut from 100 x 75 or 75 x 50 sawn timbers for head, posts and mullion and from 150 x 75 to 100 x 63 for sill and transom. The sawn timbers are planed smooth (wrought) and this planing reduces their size by about 5 by processing two opposed faces.

It is usual to specify the sizes of timber for wood windows, doors, frames and other joinery as being Ex. 100 x 75 for example. The description "Ex" denoting that the member is to be cut from a rough sawn timber size 100 x 75 which after being planed on all four faces would be 95 x 70 finished size.

Similarly the rails and stiles of casements and ventlights are cut from 50 x 50 or 50 x 44 sawn timbers which are planed (wrought) and whose finished sizes are therefore 45 x 45 or 39.

The members of the frame, casements and ventlights are joined with mortice and tenon or combed joints.

Figs. 65 and 66 illustrate the arrangement of the parts of two typical casement windows.

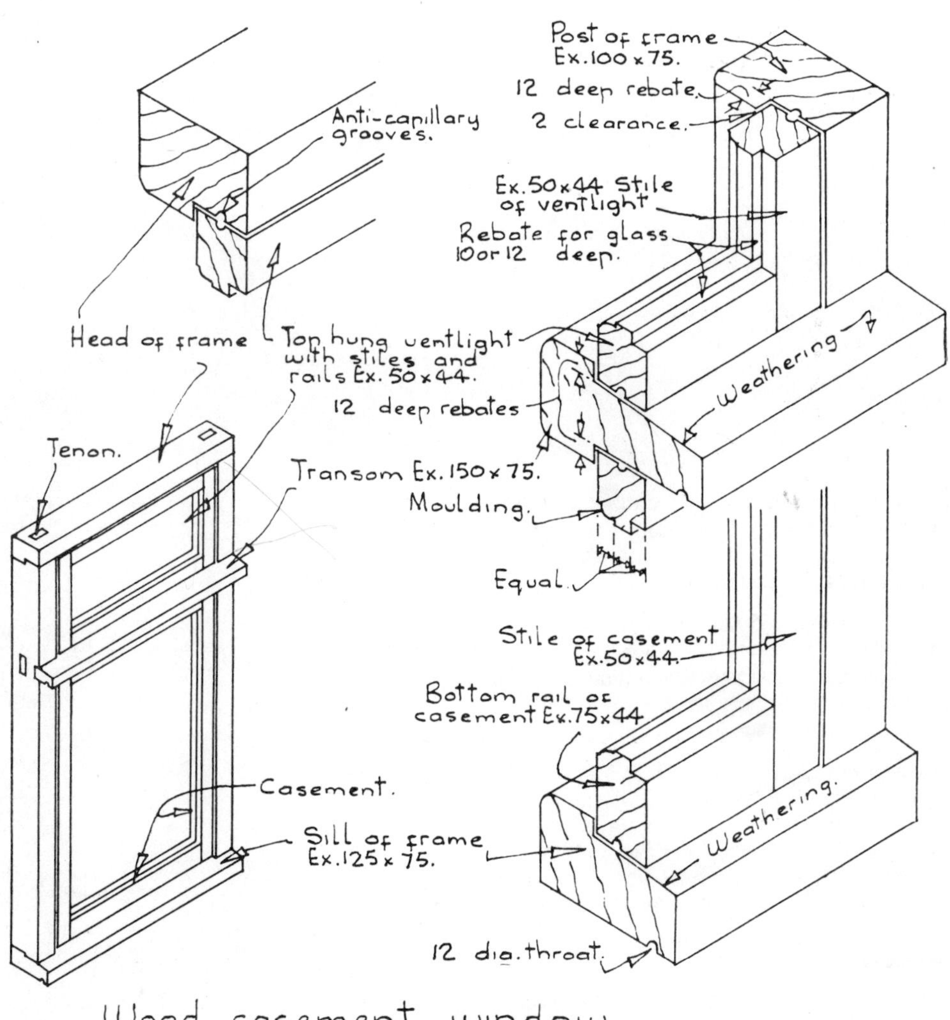


Fig. 65

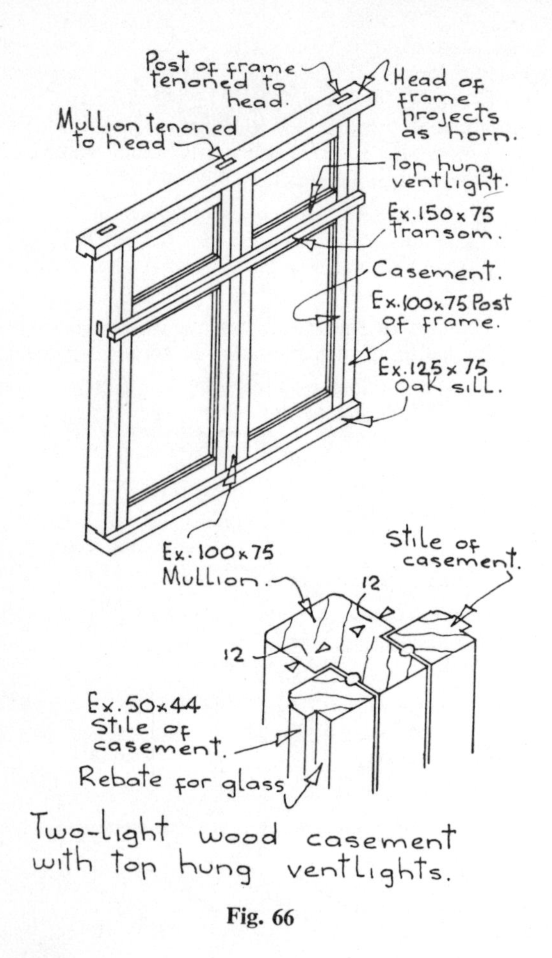


Fig. 66

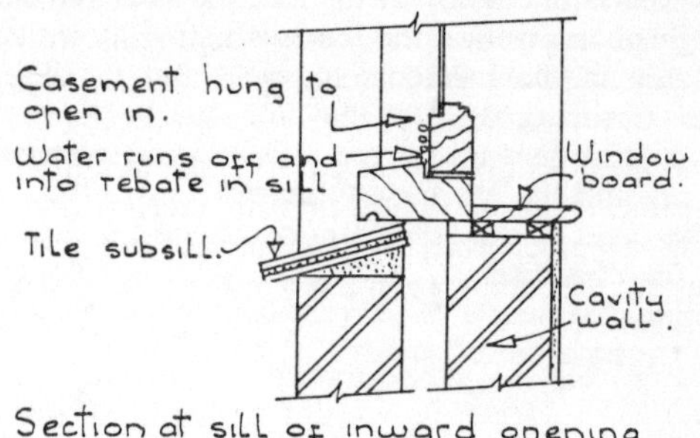


Fig. 67

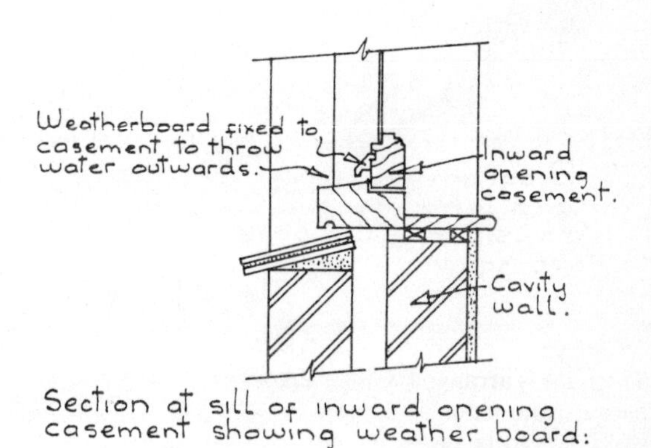


Fig. 68

It will be seen that casements and ventlights fit into rebates cut in the members of the frame. The rebates are usually 12 deep and prevent wind and rain driving in between the casement and frame.

It will be seen from Fig. 65 that semi-circular grooves are cut in the rebates in the frame and on the edges of casement and ventlight. These are anti-capillary grooves and serve to prevent rainwater finding its way between casement and frame by capillary action.

If the casements were arranged to open in and were set in a rebate in the window frame rain would run into the rebate and behind the edge of the casement at sill level as illustrated in Fig. 67. To prevent this an additional piece of timber would have to be fixed to the bottom rail of the casement as shown in Fig. 68.

**Mortice and tenon joint**

The word mortice describes a rectangular hole cut in one timber into which a tenon cut on the end of another timber can fit, as illustrated in Fig. 69.

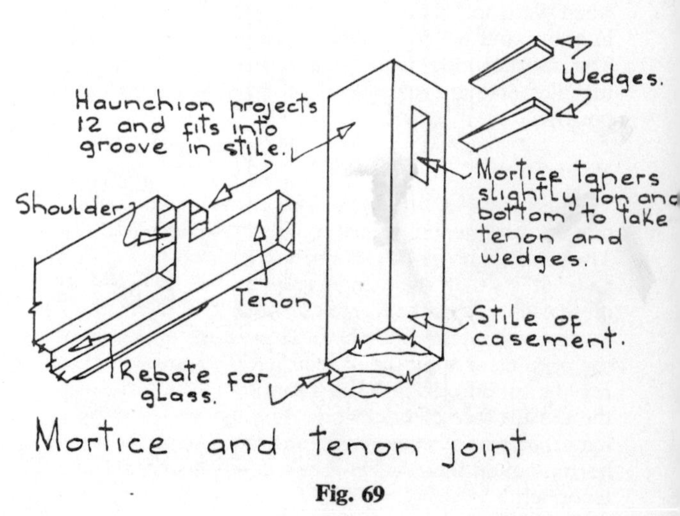


Fig. 69

If the tenon fits tightly in the mortice and it is wedged and glued in position the joint will strongly resist any tendency of the members to move out of the right angled positions in which they are joined. The strength of this joint depends on the resistance of the tenon to bending and the bearing of the shoulders of the tenon on the faces of the other timber around its mortice. Fig. 70 illustrates in section what occurs when, for example, the weight of the casement tends to cause the joint to move out of square.

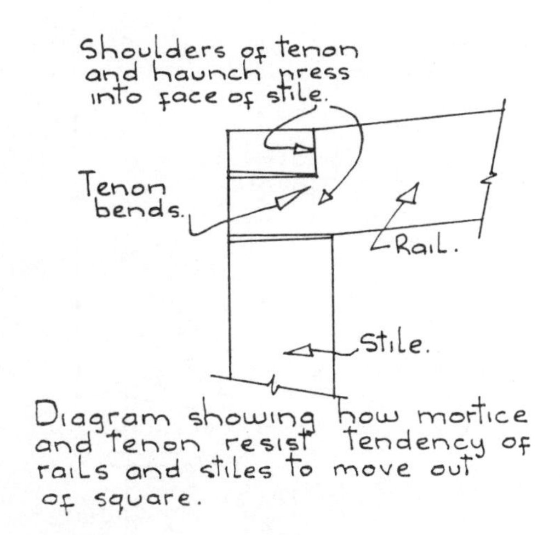


Fig. 70

The joint is arranged with as deep a tenon as practicable because the tenon is more likely to give way than the shoulders to be crushed.

The thickness of a tenon is usually ⅓rd of that of the material from which it is cut and its depth up to 5 times the thickness of the tenon.

The tenon is cut with parallel faces and the top and bottom of the mortice are cut on a slight slope so that when the tenon is in the mortice, wedges can be driven in above and below the tenon to wedge it in position. The mortice and tenon are coated with glue, put together, cramped up tightly and the wedges driven into position.

**Window frame**

The members of the frame are joined with wedged mortice and tenon joints as illustrated in Fig. 71. The posts (jambs) of the frame are tenoned to the head and sill so that their ends may project some 40 or more each side of the frame as horns. These projecting horns can be built into the brickwork in the jambs of the opening as a means of securing the frame, or they may be cut off on site if the frame is built in flush with the outside face of brickwork. The reason for using a haunched tenon between post and head is so that if the horn is cut off there will still be a complete mortice and tenon left.

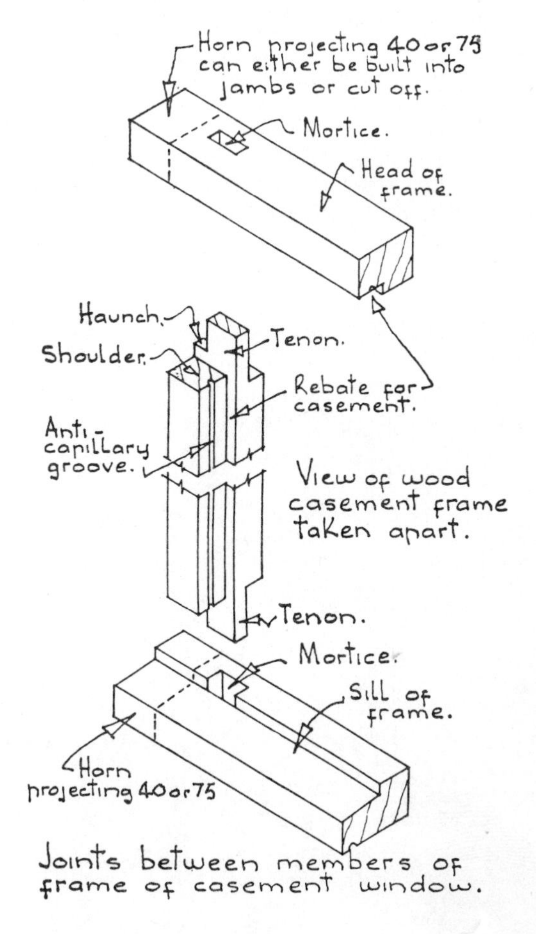


Fig. 71

It will be seen from Fig. 71 that one face of the tenons is cut in line with the rebate for the casement. It is usual practice in joinery to cut one or both faces of tenons in line with any rebates or mouldings to keep the number of faces cut across the grain to a minimum. The mortice and tenon joints are put together in glue, cramped up and wedged.

When there is a transom in the frame it is joined to the posts by means of tenons fitted and wedged to mortices. Mullions are joined to head and sill with tenons wedged to mortices and to transom with stub tenons fitted to a through mortice. A stub tenon is one which does not go right through the timber into which it is fitted.

Window frames are usually built in and secured with "L" shaped galvanised steel lugs. The lugs are 150 by 75 by 40 and the 75 arm is screwed to the back of the frame and the other one built into a horizontal joint in the jambs. One lug is used for every 300 or 450 of height of window each side of the frame.

**Casement:** The four members of a casement are two stiles, top rail and bottom rail. The stiles and top rail are cut from 50 by 44 timber and the bottom rail from 75 by 44 timber. The stiles and rails are rebated for glass and putty and rounded or moulded on their inside edges for appearance sake. Fig. 72 is an illustration of a casement. The rails are tenoned to mortices in the stiles and put together in glue, cramped up tight and wedged. Fig. 73 illustrates these joints taken apart.

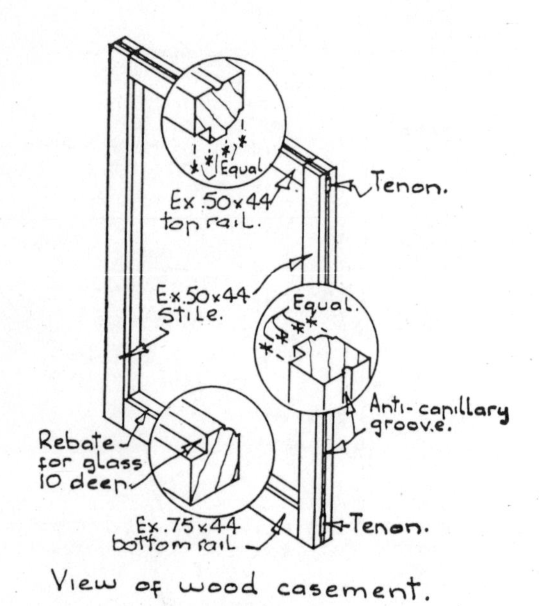


Fig. 72

The tenons are cut with their faces in line with the rebate for glass and the moulding for the reason previously given. Obviously the tenons on the rails cannot be as deep as the rails if they are to fit into an enclosing mortice and their depth is usually about half the depth of the rails on which they are cut.

**Ventlight:** The four members of a ventlight are cut from timber the same size as the stiles of the casement below and are rebated, moulded and joined with mortice and tenon joints, similar to those for casement members.

**Hinges and fastenings:** Casements and ventlights are hung on pairs of pressed steel butts screwed to frame and casement or ventlight. Cockspur fasteners and peg stays, similar to those used with metal casements, are fixed to the casements, and peg stays to ventlights.

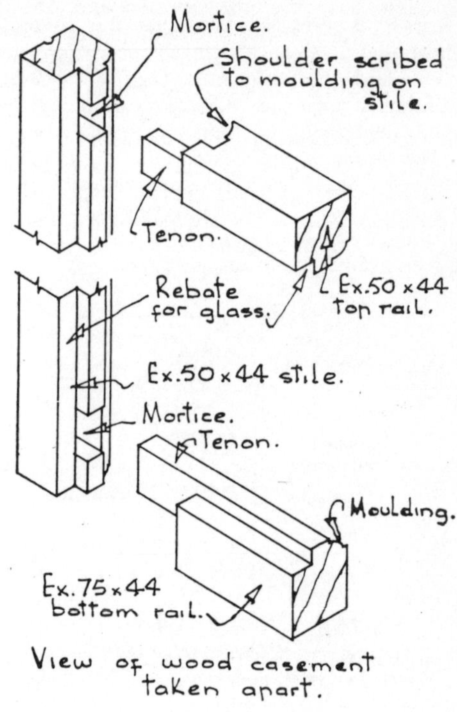


Fig. 73

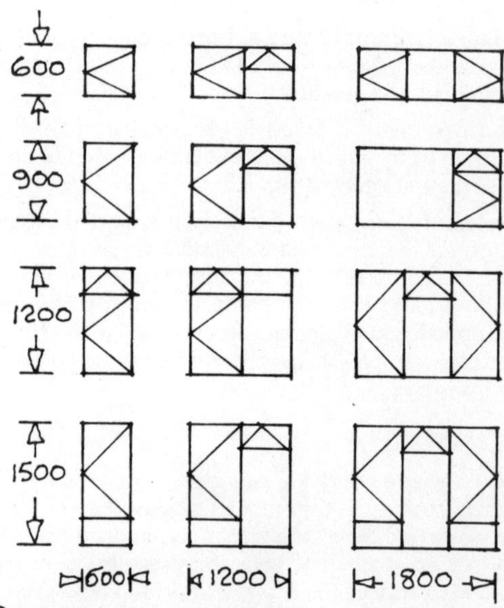


Fig. 74

**Standard wood casements**

The British Woodwork Manufacturers Association have proposed a range of demensionally co-ordinated metric standard wood casements. The frames, casements and ventlights are made from standard sections of softwood timbers put together in a variety of standard sizes. The windows are designed to have casements roughly 600 wide in various combinations with dead-lights and ventlights. Fig. 74 is an illustration of some typical standardwood casements designed to fill dimensionally co-ordinated basic spaces.

The casements and ventlights are cut so that their edges lip over the outside faces of the frame by means of a rebate in their edges. These lipped edges are in addition to the rebate in the frame so that there are two checks to the entry of wind and rain between casement and frame. The standard sections used in these windows are illustrated in Figs. 75 and 76. The members of the frame and those of the casements and ventlights are joined either with mortice and tenon joints or with combed joints.

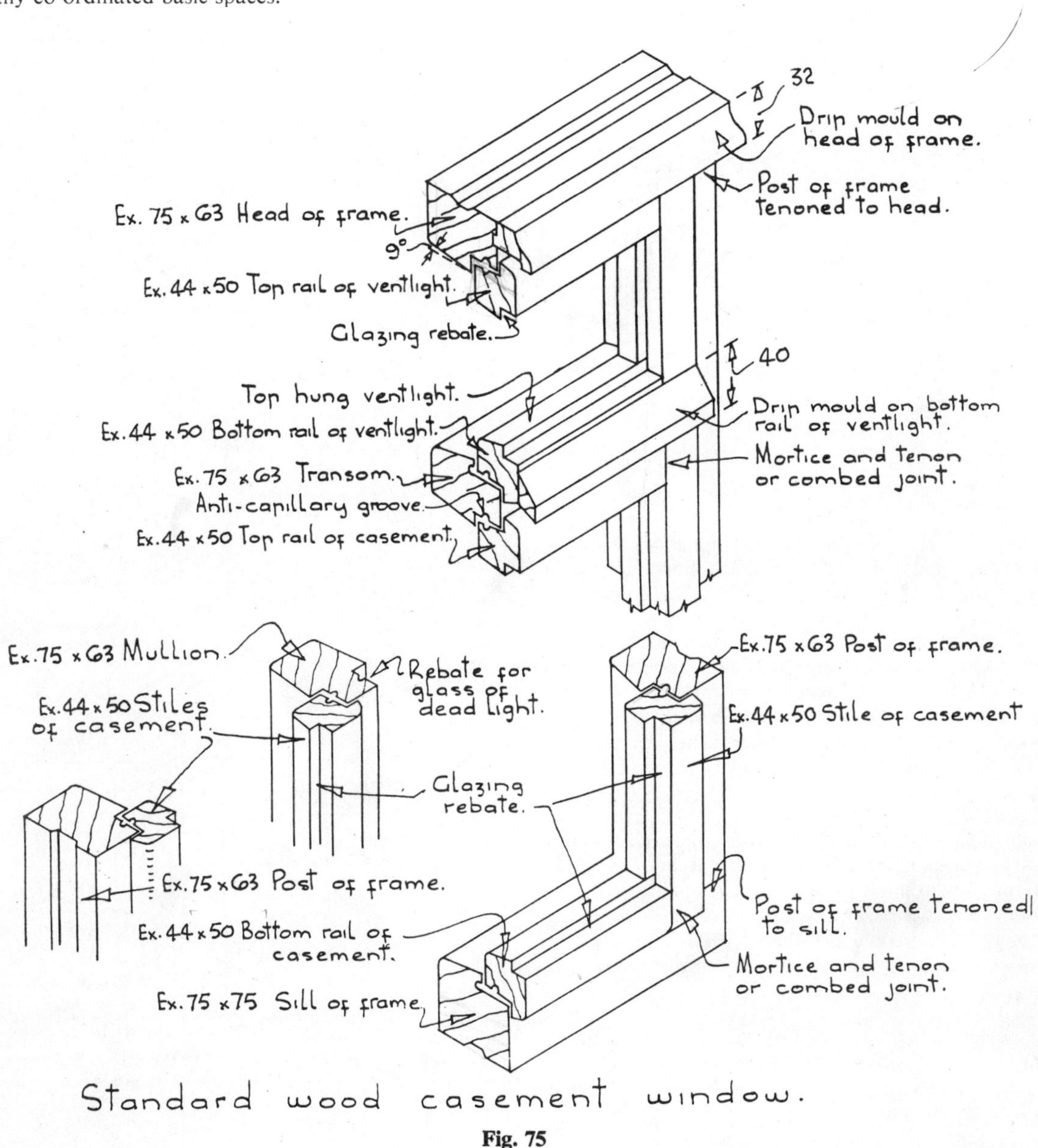


**Fig. 75**

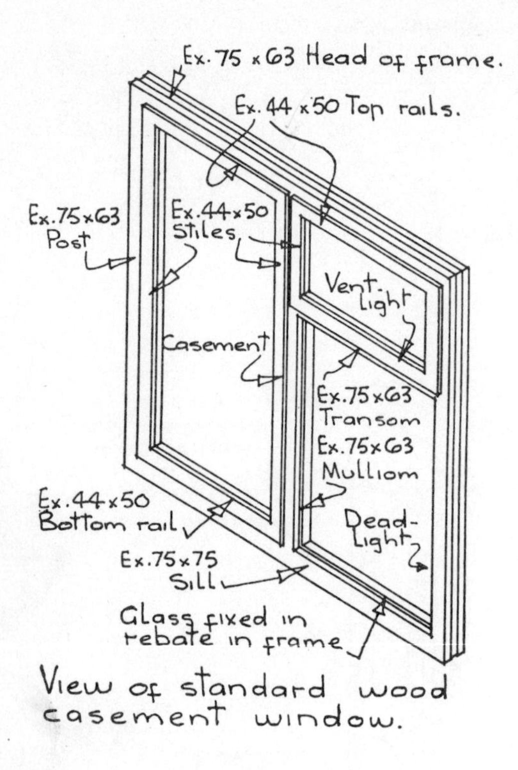


Fig. 76

**Joinery:** The craft of accurately cutting and joining the timber members of windows and doors is termed joinery and those who practise it are called joiners.

For centuries the joiners craft was executed with hand tools used for preparing, cutting and assembling timbers. A mortice and tenon joint can readily be cut and made by skilled joiners and as it very rigidly joins timbers it was the joint always used in framing the members of windows and doors up to some fifty years ago.

During the present century woodworking machinery has been increasingly used to prepare, cut and assemble windows and doors so that today the majority of windows and doors are machine made. Woodworking machinery has made it possible to produce a range of standard windows and doors at comparatively low prices.

The skilled joiner can quickly cut and assemble a mortice and tenon joint but the time taken by machinery to cut and assemble this joint is greater than that required to cut and assemble a combed or dowelled joint. In consequence mortice and tenon joints are much less used for framing window and door members than they were.

**Combed joint**

This joint is commonly used to join the members of standard wood casements and ventlights because it can economically be cut and assembled by wood working machinery. Fig. 77 is an illustration of a typical combed joint. There must not be less than two tongues on the ends of members to be joined and no tongue should be less than 7 thick.

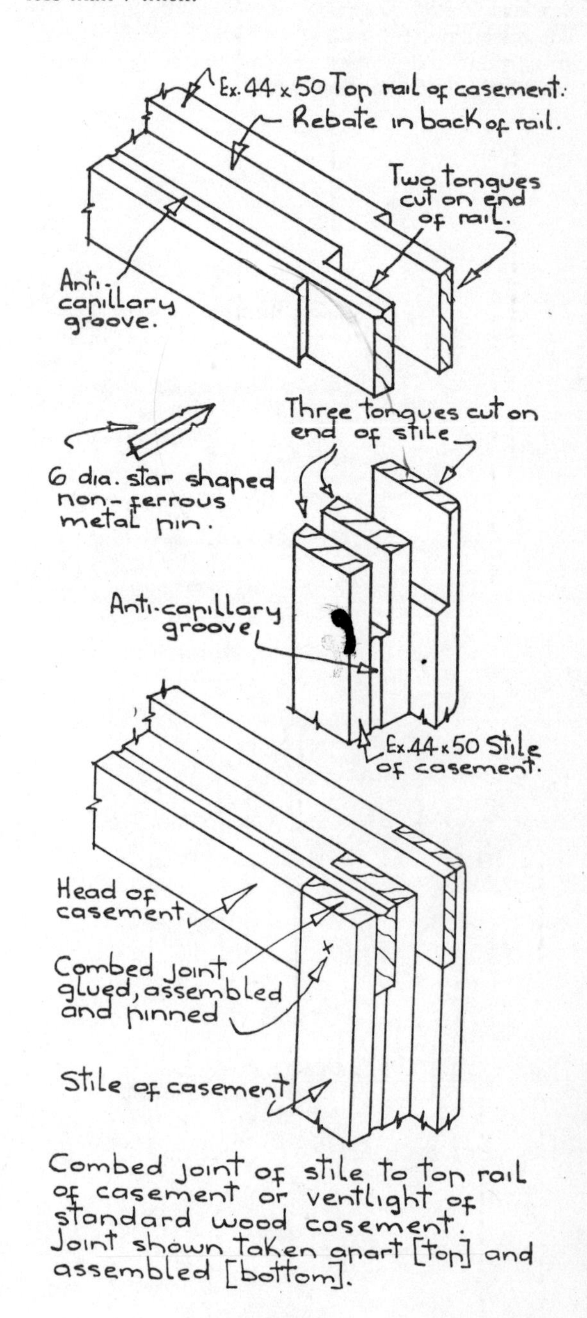


Fig. 77

This joint is put together with the faces of the combs coated with glue and a wood or metal dowel pin is driven through the joint. If accurately cut and properly put together this joint is sufficiently strong for casements, ventlights and window frame.

Because standard wood casement windows are made in standard sizes their cost compares very favourably with standard metal casement windows of similar size. These windows are usually well made, are of pleasant proportions and if they are well maintained will be trouble free for very many years.

**Pivoted sash windows**

Fig. 78 is an illustration of a horizontally pivoted sash window. The frame is cut from timbers size 75 x 50 or 100 x 50 joined with mortice and tenon joints. The frame is built in and secured with "L" shaped building-in lugs as previously described. The sash consists of two stiles and top and bottom rails which are joined with mortice and tenon joints similar to those used for casements.

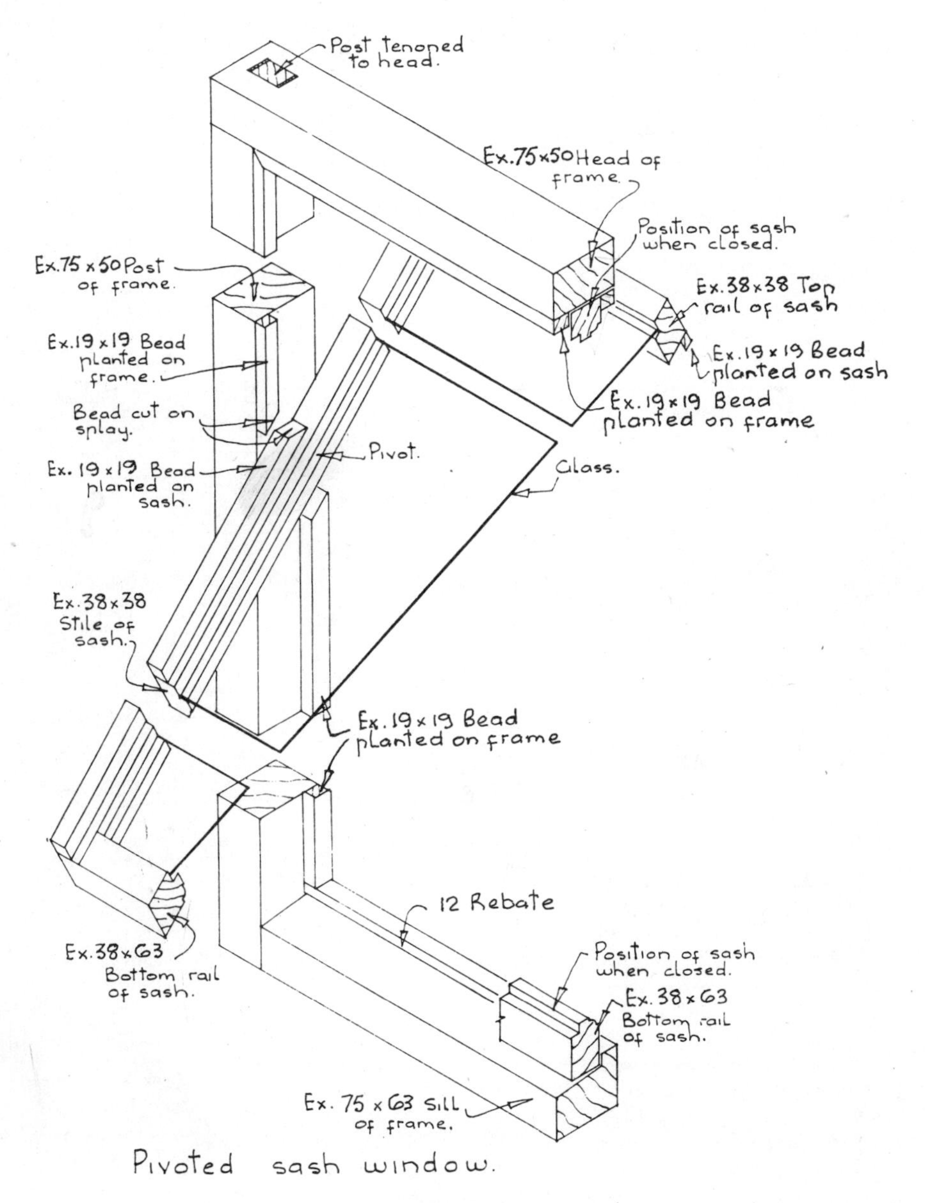


**Fig. 78**

Timber beads are planted (nailed) to the sash and the frame so that the top of the sash, which opens in, fits behind beads planted on the outside of the frame, when closed and the bottom of the sash to beads planted on the inside of the frame when closed. Beads are nailed to the sash, outside below the pivots, and inside above the pivots so that when the sash is closed there are beads all round the frame both sides. These latter beads are fixed for the sake of appearance only.

Beads are used in this type of window to prevent wind and rain getting between sash and frame, rather than rebates cut in the frame which would be expensive to cut.

The sash is pivoted with its bottom opening out so that when it is open rain falling on it runs down the glass and away from the building. The maximum economical size of horizontally pivoted sash is 1·5 x 1·5. A larger sash would project too far into the room when open and would require thick members to resist the tendency of the sash to bend over the pivots when open, due to its own weight.

A better method of fixing beads for this type of window is illustrated in Fig. 79. It will be seen that only one set of beads is used and planted (nailed) to the frame above the pivots, and to the edges of the sash below the pivots. The appearance of this type of pivoted sash window is neater than that with beads planted on the face of the frame and sash and there is less likelihood of the beads coming away due to rain soaking between them and the frame and sash.

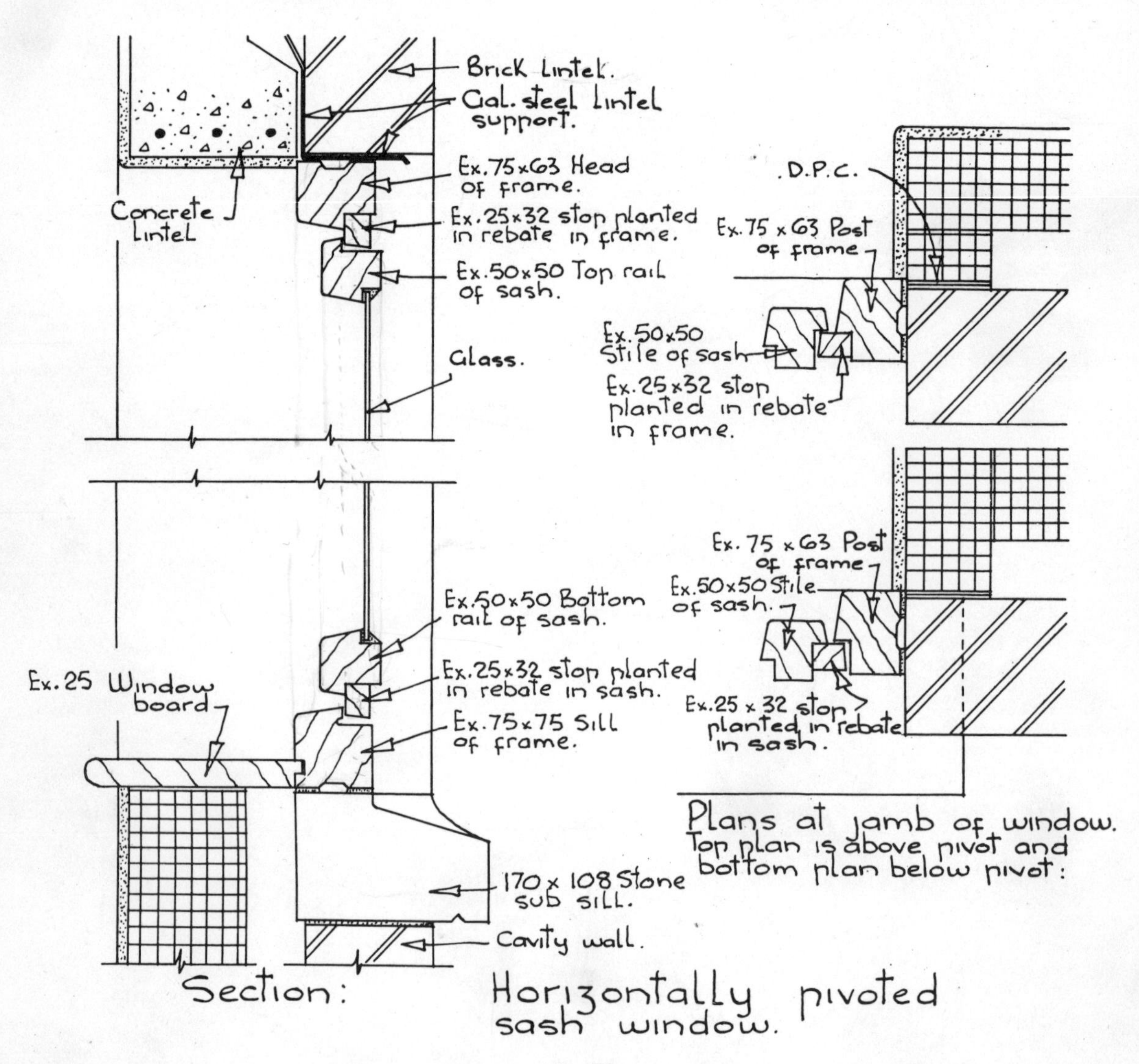


Fig. 79

**Double glazed pivoted sash windows**

Of recent years double glazed windows have become popular. Double glazing consists of two sheets of glass fixed a short distance apart. The purpose of using double glazing instead of the conventional single sheet of glass is to reduce losses or gains of heat through windows. The reduction in transfer of heat is achieved mainly by virtue of air trapped between the two sheets of glass. Pivoted sash windows are particularly suitable for double glazing because they consist of one large unobstructed sash which is cheaper to fit with double glazing than, say, a number of smaller casements.

Fig. 80 is an illustration of one of the horizontally pivoted double glazed windows being manufactured. The maximum economical size of sash that can be used is 1·5 wide by 1·5 high. The frame is of conventional construction, cut from rectangular timber sections, put together with mortice and tenon or combed joints and the sash closes to beads planted on the frame and sash as previously explained. The opening part of the window consists of two sashes, an inner sash suspended on pivots fixed to the frame and an outer sash side-hinged to the inner sash. Each sash is glazed with a separate sheet of glass. Two sashes are used so that the two sheets of glass can be cleaned on both sides from inside the building. Fig. 81 illustrates how the sash can be turned through 180 degrees so that the outer side-hinged sash can be opened into the building for cleaning the

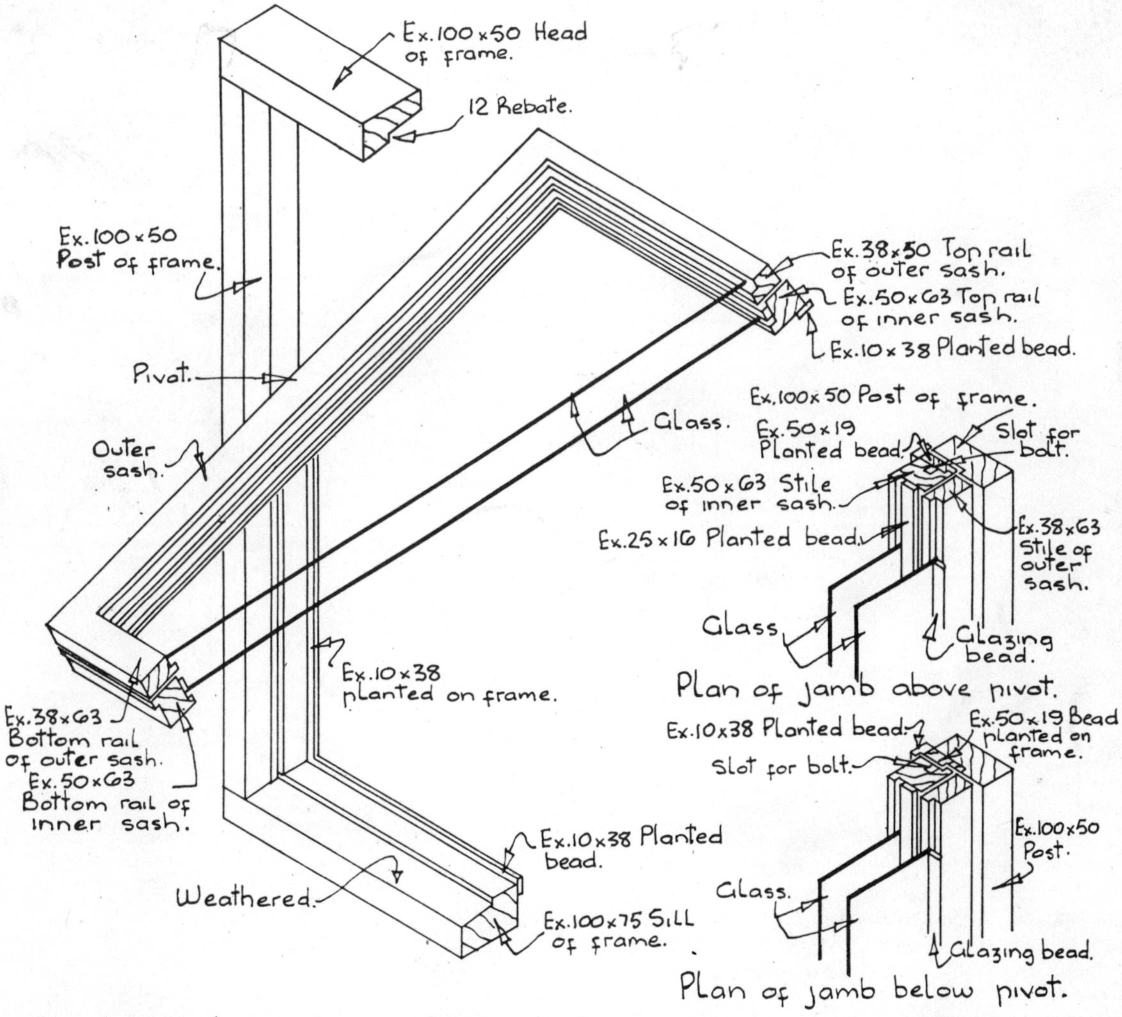


**Fig. 80**

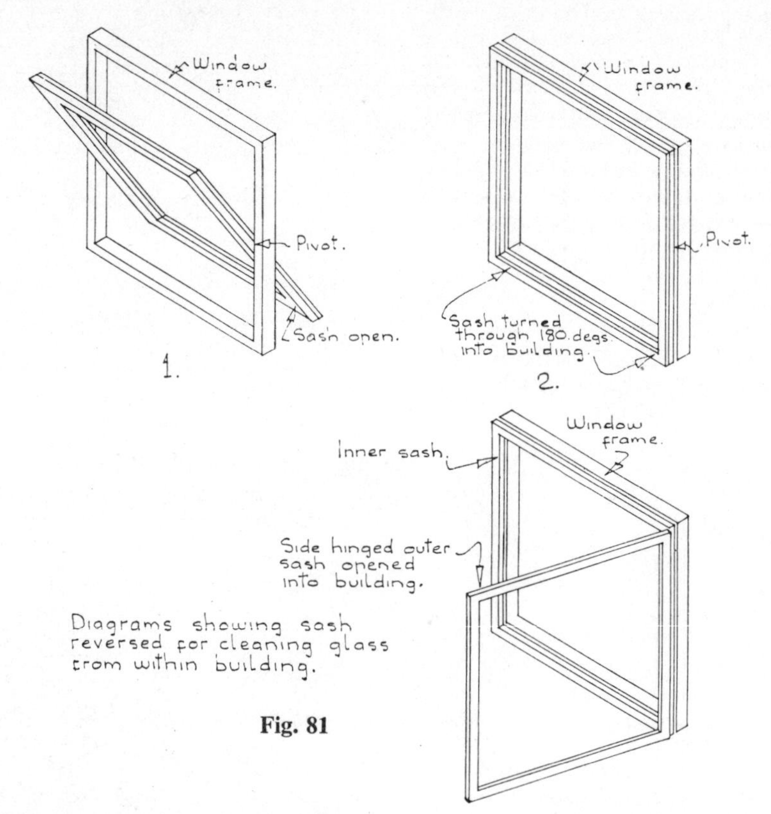


Fig. 81

two sheets of glass. The two sashes are normally locked together. The sash can be locked by means of a lever which operates espanolite bolts housed in the stiles of the inner sash.

The window is maintained in a selected open position by means of the friction pivots on which it is hung. At present double glazed pivoted windows are considerably more expensive than other types of window of similar size.

**Vertically pivoted sash window**

In common with horizontally pivoted windows this type of window can be made with one large unobstructed sash. The sash is hung in pivots fixed to the head and sill of the frame so that the major part of the width of the sash opens out of the building for the reason previously explained. But vertically pivoted sash windows are not as popular at present as horizontally pivoted ones because they suffer the following disadvantages. When the sash is open it may act like a sail in wind and cause uncomfortable eddies of air around itself and inside rooms. Even in the lightest wind it is often impossible to prevent this. The part of the sash that opens into the building obstructs movement around the window and may be a danger to the unwary person moving near the window. Vertically pivoted windows are generally used in lieu

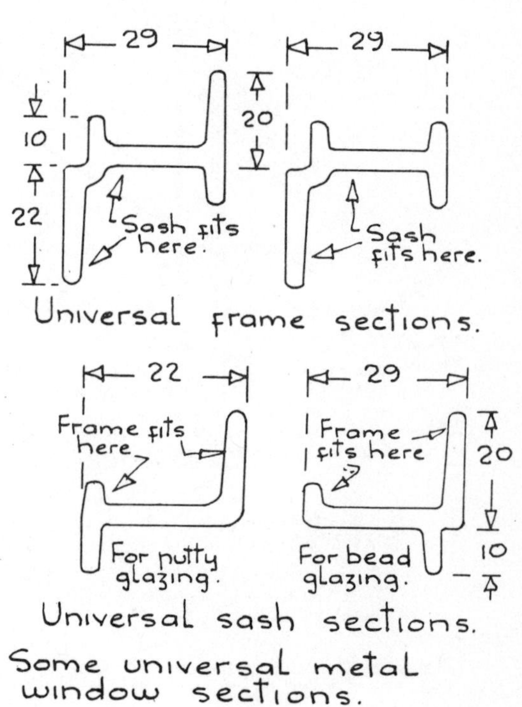


Fig. 82

of casements because their sashes can be made wider than a casement without being liable to sink out of shape.

Vertically pivoted timber windows are constructed in the same way as horizontally pivoted windows with solid frames and sashes closing to beads fixed to the frames and sashes. The beads may be fixed to the face of the sash and frame or fixed in the edges of the sash and on the frame as shown in Figs. 78 and 80.

Vertically pivoted metal windows are used principally in schools and factories. The frames and sashes are made from "Universal" mild steel sections cut to length and welded together at the angles of frame and sash. Fig. 82 is an illustration of some typical "Universal" mild steel sections used.

Fig. 83 illustrates a typical vertically pivoted metal window.

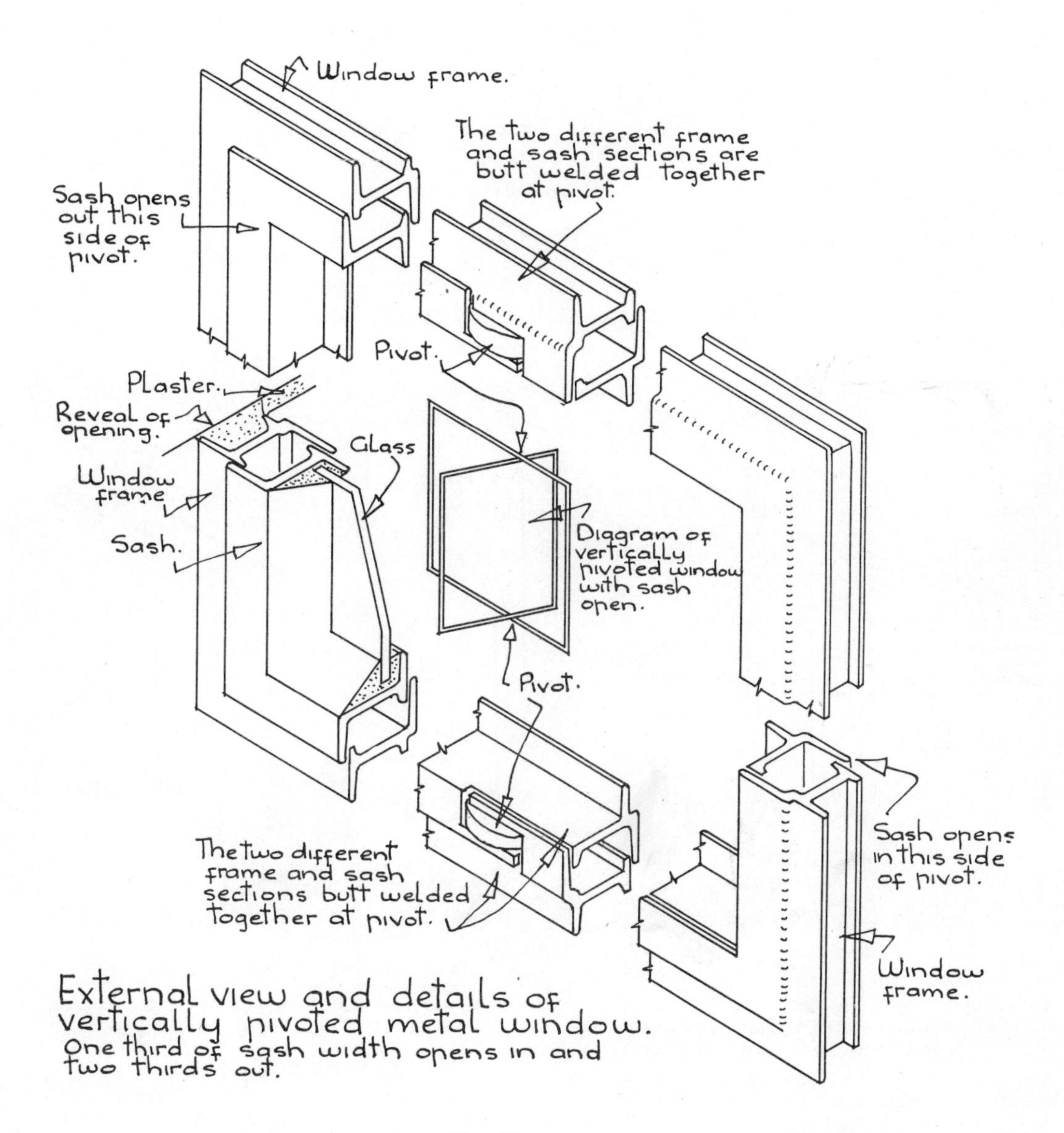


Fig. 83

**Vertically sliding sash window (Double hung sash window).**

**Cased frame:** Up to about twenty years ago the sashes in this type of window were usually suspended on cords running over pulleys and attached to counterbalance weights inside the window frame which was made in the form of a narrow box (case) to accommodate them.

The four parts of the frame are two jambs, head and sill.

**The jambs,** are cased from three thin members tongued, grooved and glued together as illustrated in Fig. 84. The pulley stile, in which the pulley supporting the sash cords is fixed, is usually 7 thicker than outer and inner linings because it carries the weight of the sashes and counterbalance weights. A thin strip of wood, the parting slip, is suspended inside the cased jambs to separate the weights of top and bottom sashes. A strip of plywood or hardboard is nailed across the back of the linings as back lining, to prevent the weights fouling the reveals of the opening.

The size of the cased jambs depends on the thickness of the sashes. If the sashes are cut from 38 thick

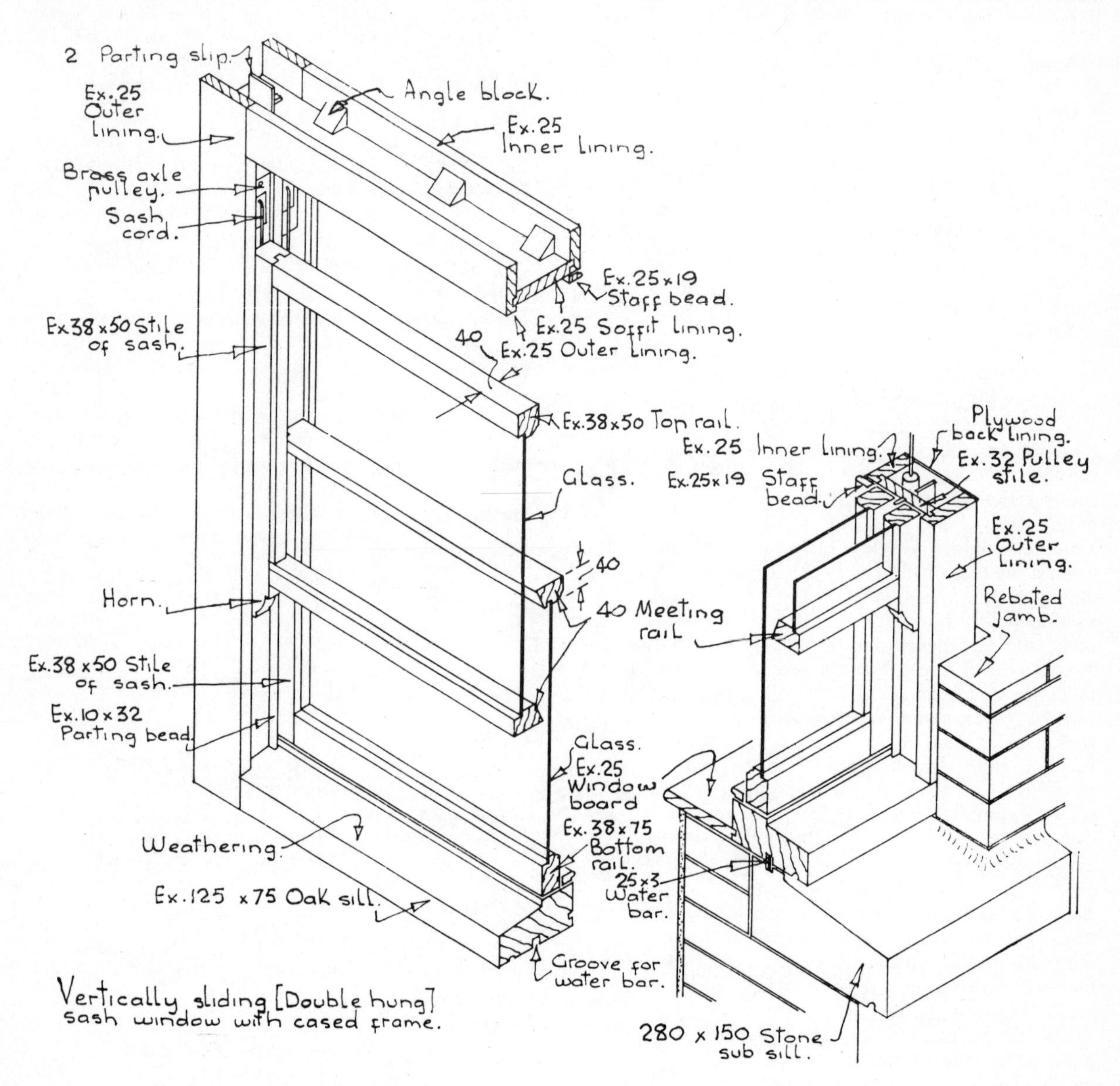


**Fig. 84**

timbers the inside of the cased jamb is usually 85 wide and 50 deep and with sashes cut from 44 thick timbers 105 and 50 deep.

**The head** of the cased frame may be constructed from three thin timbers put together with glued tongued and grooved joints as illustrated in Fig. 84, or as a solid rectangular section of timber 38 deep and as wide as the cased jambs overall.

**The sill** of the frame is cut from a solid section of oak 75 deep and as wide as the cased jambs overall. The sill is weathered and sunk on its top surface. The word weathered denotes that the top surface of the sill is cut to slope outwards to throw rainwater off. The sinking is a shallow rebate some 6 deep in line with the face of the lower sash. It serves to prevent rain-water being blown between the sill and the bottom edge of the lower sash.

A throat and a groove are cut on the bed or underside of the sill. The throat is a 12 wide groove which serves the same purpose as the throat on the underside of a coping stone (Vol. 1). The groove is cut to take a 25 x 3 galvanised steel or wrought iron water bar which is set partly into the groove in the oak sill and partly into a similar groove in the stone sub sill (see Chapter 3).

The water bar is bedded in mastic in the oak sill and in cement in the stone sub sill and serves as a barrier to the entry of water between the two sills.

The head of the frame is cut to fit between the linings of the jamb and the pulley stile is tongued to a groove in the soffit. The sill is cut to fit between the lining of the jambs and the pulley stile is wedged into a groove in the sill.

The outer linings of the jambs and head project 12 beyond the face of the pulley stile to act as a guide for the top sash.

Parting beads 10 x 32 are set into grooves in the pully stiles to separate and act as guides for the sashes. A staff bead is screwed to the edge of the inner linings of jambs and head and to the sill to act as a guide for the lower sash. When a sash cord breaks the staff bead is removed to lift the bottom sash into the building and the parting bead is also taken out if the top sash has to be taken out. The bead attached to the sill is sometimes made 38 deep so that the lower sash can be lifted slightly to allow air to enter between the meeting rails without causing direct draughts of air.

To renew sash cords the sashes are lifted out of the frame into the room. Small traps cut in the bottom of the pulley stiles, called pockets, are taken out so that the new sash cords can be attached to the weights inside the cased jambs.

Brass axle pulleys, two to each sash, are fixed in the pulley stiles as illustrated in Fig. 84.

**Sashes** are usually of the same depth so that the meeting rails come together in the middle of the height of the windows when they are closed.

The stiles and top rail of sash are cut from 38 or 44 thick by 50 deep timbers, rebated for glass and moulded inside. Meeting rails are cut from 63 x 38 timbers rebated and splayed to meet as shown in Fig. 84 and rebated for glass. The bottom rail of bottom sash is cut from timber 38 or 44 thick by 63 or 75 deep rebated for glass and moulded.

The top rail of top sash and bottom rail of bottom sash are tenoned, wedged and glued to mortices in the stiles as described for timber casements.

So that tenons the full depth of the thin meeting rails can be cut, it is usual to cut the stiles of sashes so that they project beyond the meeting rails as horns. The projecting horns are moulded for appearance sake as shown in Fig. 85.

**Sash cord** is made from twisted or braided flax and cotton in cord some 6 thick. In time the cord frays and breaks and needs fairly frequent renewal. Cord made from a mixture of nylon fibre and flax is also made for hanging sashes. It is more expensive than flax cord but has a longer life. Heavy sashes are sometimes hung on copper chain used in lieu of cords. Although expensive the chain does not easily break.

The sashes are locked in the closed position by means of a sash fastener which consists of a pivoted bar attached to a plate fixed to one meeting stile and which closes to a catch fixed to the other meeting rail. To facilitate opening and closing sashes it is usual to fit sash lifts to the bottom rail of the lower sash and cords and pulleys for raising and lowering the top sash. The lifting cords operate over pulleys fixed to the head of the window frame and are attached to the stiles of the sash, and the cords for lowering the sash are fixed directly to the stiles.

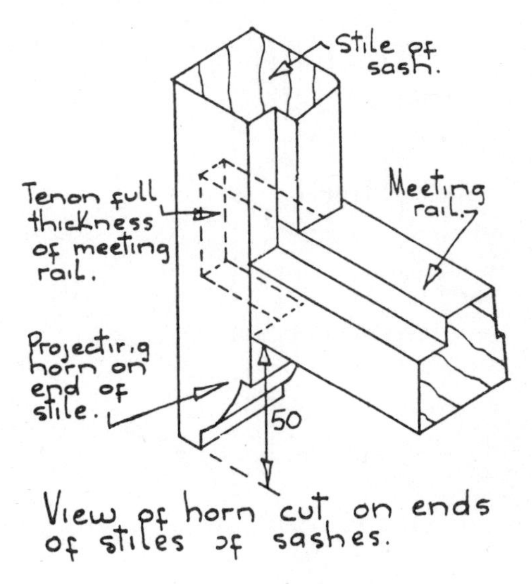


Fig. 85

**Vertically sliding sash window with solid frame**

For the past thirty years a spiral sash balance has been manufactured for use instead of cords and counter-weights to balance vertically sliding sashes. The balance does away with the need for a cased frame to the window and so simplifies its construction. The spiral balances consist of a metal tube inside which a spiral spring is fixed at one end. Fixed to the other end of the spring is a metal cap with a slot in it through which a twisted metal bar runs. The tube is fixed to the window frame and the twisted bar to the bottom of the sash, two balances being used for each sash. As the sash is raised or lowered the twisted bar tensions the springs in the tubes. The sash is supported (balanced) by the tension of the springs in the balances and can be raised or lowered with little effort.

Fig. 87 is an illustration of one sash balance.

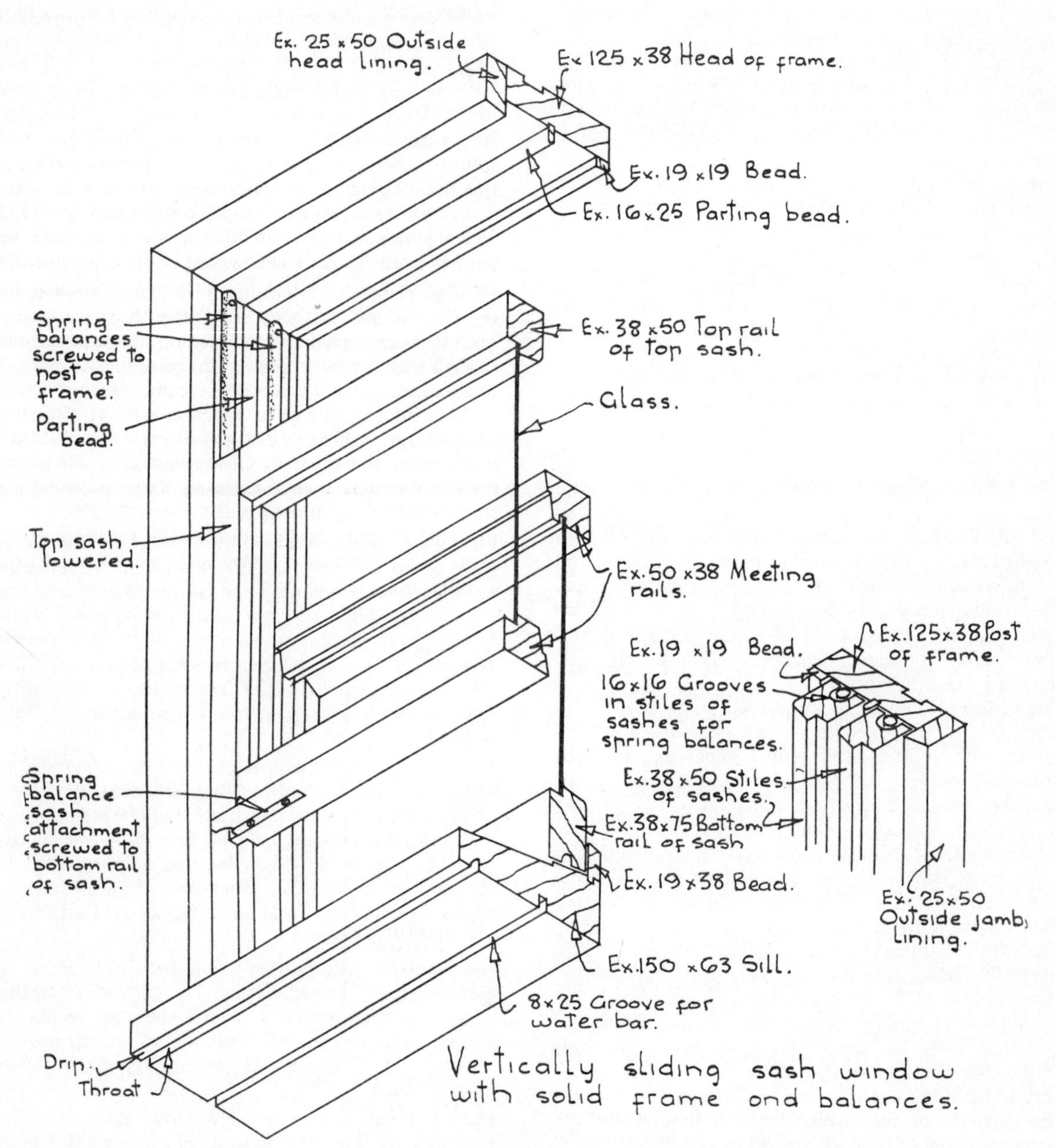


Fig. 86

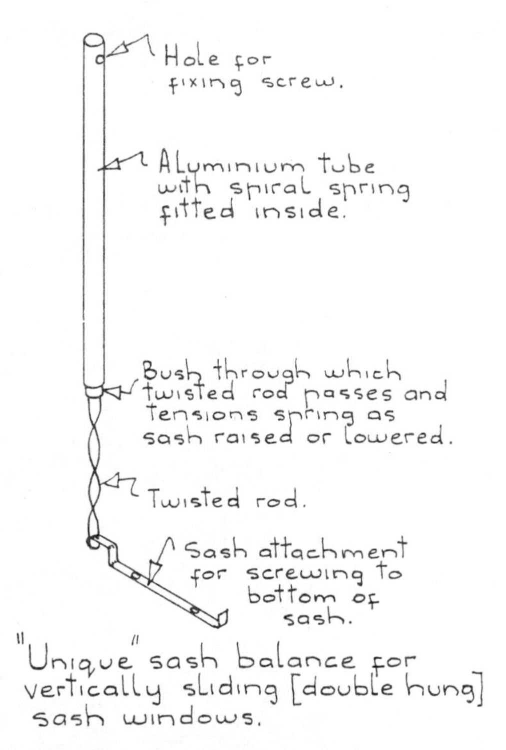


**Fig. 87**

The window frame is constructed from four solid rectangular sections of timber, two posts (jambs), head and sill. The posts are joined to the head and sill with combed joints with three combs on each end of head and sill into which two combs on the ends of posts are fitted. The joints are glued and pinned with 6 wood or metal dowels. These combed joints are of similar construction to those described for use with the standard wood casement.

The sashes are similar to those used for windows with cased frames. The members of the sashes are put together with mortice and tenon joints as described previously or with combed joints similar to the joint described for standard wood casements.

A range of vertically sliding windows with cased or solid frames, cut from standard sections put together in a variety of standard sizes, is manufactured. By virtue of standardisation of sizes these windows are cheaper than similar purpose made ones.

Fig. 86 illustrates the construction of a standard vertically sliding sash window with spring balances.

One method of fixing the frame of vertically sliding windows is to build it into rebated jambs. The jambs are built with $\frac{1}{2}$B. deep rebates so that the whole of the frame is set into the rebate and only the sashes appear from the outside. This is illustrated in Fig. 84. The purpose of this arrangement is to give the appearance, from outside, of a window consisting almost entirely of glass.

The other method of building in these windows is to set them inside square jambs with the outside of the frame some 25 or more back from the external face of wall. A wood architrave is then planted around the frame.

**GLASS**

Is made by heating soda, lime and sand to a temperature at which they melt and fuse. The molten glass is either drawn continuously as clear sheet glass or rolled out and polished both sides as polished plate glass.

**Clear sheet glass:** Clear sheet glass is drawn from a bath of molten glass. The usual thicknesses of sheet glass are 3, 4 and 5.

Because this glass is not exactly flat it causes some optical distortion of images seen through it. The glass is graded as O.Q. (ordinary glazing quality), S.Q. (selected glazing quality), S.S.Q. (special selected quality), the latter qualities being selected as more nearly flat and free from imperfections than the O.Q. The ordinary quality is used for windows, the S.Q. for glazing picture frames and special selected for mirrors.

**Polished plate** glass is produced by a continuous drawing process and the glass is polished both sides until both surfaces are flat and smooth and the sheet uniform in thickness. Polished plate is manufactured in various thicknesses, some thicknesses commonly used being 3, 5, 6 and 10. This glass causes no optical distortion of images seen through it. Polished plate glass is about three times the price of clear sheet glass of similar thickness and is used for glazing large windows for shop windows, counter tops and mirrors.

**Float glass** is produced by continuously running molten glass on to a base of molten lead. The molten glass solidifies as it passes over and off the molten lead and emerges as a sheet of glass both faces of which are so smooth as to require no polishing. There is at present very little difference between the cost of plate and float glass and they are used for like purposes in building. The thickness of float glass commonly used is 3, 5, 6 and 10.

**Translucent glass (obscured glass)** is manufactured by impressing one or both surfaces of glass with some pattern or texture so that images cannot be seen through the glass. Most obscured glass is figured on one surface and the other is flat. This glass is manufactured by rolling or casting. The various types of figure impressed on the glass are described by such names as "Cathedral", "Arctic", "Hammered".

**Glazing:** The operation of fixing glass in windows or glazed doors is termed glazing. The glass is cut so that there is a clearance of 2 around the glass inside the glazing rebate. The two methods of fixing glass in windows and doors are (1) with putty and (2) with glazing beads.

**Putty:** For glazing to wood windows is composed of powdered whiting and linseed oil. Whiting is a finely divided powder of pure natural chalk. Linseed oil is a vegetable oil extracted from the seed of the linseed plant. It is one of the so-called "drying oils" as it oxidises on exposure to air and a softish, elastic water-repellant film forms on its surface. The whiting and linseed oil are ground together until the mixture is a smooth paste, which is sold in sealed drums. Glass is bedded in putty in the glazing rebate of casements and sashes as illustrated in Fig. 88. A thin bed of putty, termed back putty, is spread inside the rebate into which the glass is pressed. To keep the glass in position while the putty is hardening glazing sprigs are tapped into the rebate as illustrated in Fig. 88. Putty is then spread in the rebate of the casement or sash and finished smooth and weathered (sloping). Putty gradually hardens over the course of a few weeks.

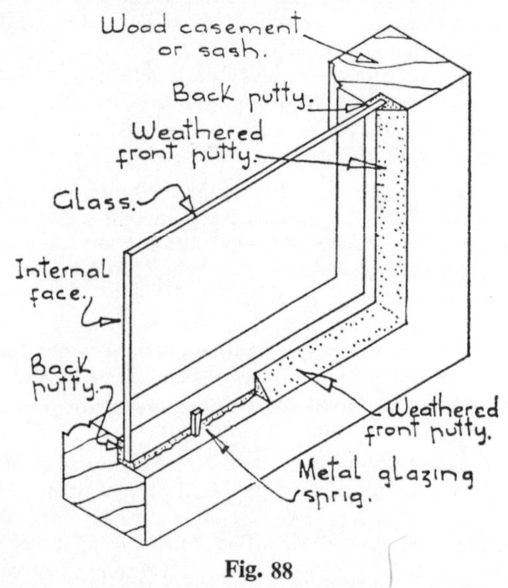


**Fig. 88**

Putty for glazing to metal windows is composed of refined vegetable drying oils mixed with finely ground chalk. The glass is bedded in back putty and secured, whilst the putty is hardening, with metal spring clips as illustrated in Fig. 89. Then weathered putties are made as described for glazing to wood.

**Glazing beads:** Windows and glazed doors are sometimes made of hardwood which, instead of being painted, is polished or varnished to show the attractive colour and grain of the wood. It would spoil the appearance of these windows and doors if glass were fixed with putty, the colour of which would contrast unpleasantly with the wood.

Glass in polished or varnished hardwood windows and doors is secured with hardwood beads fixed in a glazing rebate internally, as illustrated in Fig. 90. It will be seen that the glazing rebate is cut on the inside and is 12 deep and 25 or more wide to accommodate the bead. Narrow strips of wash leather (chamois leather) or glazing felt are wrapped around the edges of the glass in the depth of the glazing rebate. These

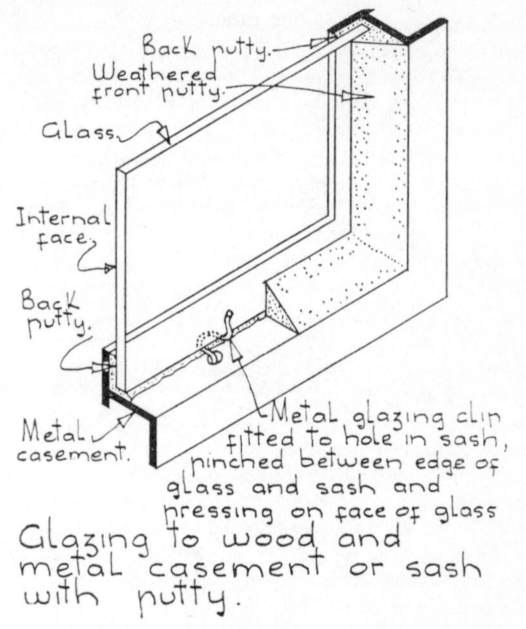


**Fig. 89**

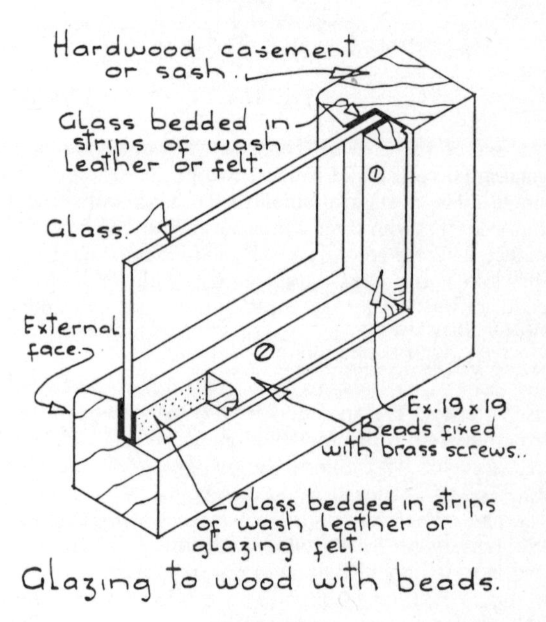


**Fig. 90**

strips cushion the glass and prevent it rattling as the window or door is opened. The glazing beads are usually of the same wood as the door or window and are secured with brass screws and eyes. Brass eyes are circular washers which prevent the screw head biting into the wood and so making an unsightly hole. The beads are mitred (cut at 45 degrees) at corners so that the moulding cut on them returns neatly round the corner.

**Double glazing:** One thin sheet of glass in a window or door is a very poor insulator against transference of heat. In an average heated building as much as 25% of all heat losses may be due to heat lost through thin glass in windows. In order to reduce the amount of heat lost through glass it is not economic to increase the thickness of glass used, as by doubling its thickness the cost of the glass is more than trebled and the heat lost through it only reduced by say 5%. The usual method of reducing heat losses is to fix two sheets of glass in a window so that they are spaced from 3 or more apart. The air trapped between the sheets considerably improves insulation. In general the further the sheets are spaced apart the better the insulation. But if two sheets of glass are fixed in a casement or sash and the air space between them is not hermetically sealed, fine dust will find its way into this space and in a short time make the glass so dirty that it cannot be seen through. In addition any moisture in the air between the sheets will condense on the inside of them in cold weather and again it will be impossible to see through the glass. Obviously then either the space between the glass must be sealed and filled with dry air, or means must be provided for cleaning the inside surfaces of the two sheets of glass.

Several firms supply double glazing consisting of two sheets of glass spaced from 3 to 12 apart, sealed around the edges and with dried cleaned air between the sheets. Fig. 91 is an illustration of typical double glazing of this type. At present this type of double glazing is very expensive compared to the cost of a single sheet of glass of similar size and in consequence has limited use. The other method of double glazing windows is to fix two sheets of glass in separate sashes which are normally locked together but can be opened for cleaning as illustrated in Fig. 80.

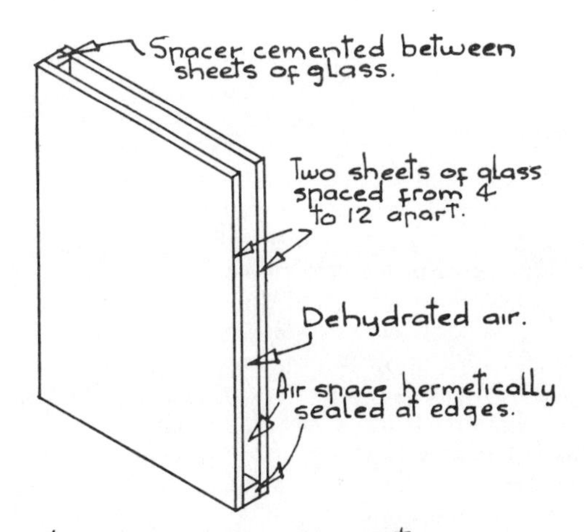


**Fig. 91**

### Reference Books and Publications

**British Standards:**
No. 544. Linseed oil putty.
No. 606. Plaited sash cord.
No. 644. Part 1. Wood casement windows.
Part 2. Wood double hung sash windows.
Part 3. Wood double hung sash and case windows.
No. 952. Glass for glazing.
No. 990. Steel windows for domestic buildings.
No. 1285. Wood surrounds for steel windows and doors.
No. 1331. Builders' hardware for housing.
No. 1422. Steel subframes, sills and window boards or metal windows.
No. 1787. Steel windows for industrial buildings.

**British Standard Code of Practice:**
CP. 3. Chapter 1 (A), Daylight; and Chapter 1(C), Ventilation.
CP. 151. Part 1. Doors and windows including frames and linings.

**Building Research Station Digests:**
No. 34. The principles of natural ventilation of buildings. (First Series).
Nos. 41 and 42. Estimating daylight in buildings.
No. 76. Integrated daylight and artificial light in buildings.

CHAPTER THREE

# WINDOW SILLS

## SfB (31) Windows: Sills: General

A window frame is usually less thick than the wall in which it is built so that, unless the frame is set flush with the outside face of the wall, there are horizontal surfaces of brickwork either side at the foot of the window. Most of the area of a window is glass which does not absorb water and rain runs off it on to the external surface below. To prevent this rain saturating the brickwork below the window, a sill is constructed. The sill may be of wood, stone, tile, brick, sheet metal or any reasonably dense material which will not readily absorb moisture. The sill is constructed to slope out from the window to throw water clear of the wall below. The sloping top surface of a sill is described as weathering, or, because of the sloping surface, it is said that the sill is weathered.

The internal surface at the bottom of a window will collect dust and may become damp from moisture which condenses on the inside face of the glass and runs down. It is usual to fix or construct an internal sill of some material which is reasonably hard and can be easily cleaned. A timber board, called a window board, is commonly used or clay or concrete tiles may be used. The internal sill is usually fixed true level.

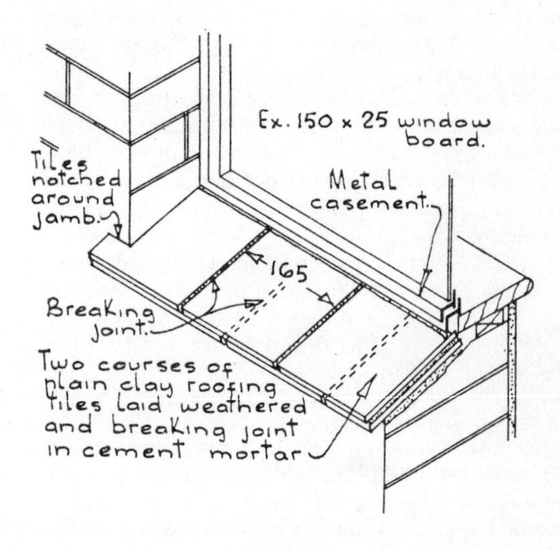


Fig. 92

### External sills

The most commonly used sills are as follows:

(1) **Two courses of plain roofing tiles, laid weathered and breaking joint in cement mortar.** The arrangement of the tiles is illustrated in Fig. 92. It will be seen that the back of the tiles is fitted under the frame and they slope outwards and overhang the wall by some 40 so that water drips off clear of the wall. If roofing tiles are used the lower course of tiles is laid so that the nibs on the tiles project and act as a drip. The butt side joints between the tiles in the top course lie over the centre of the tiles in the course below so that water cannot easily seep through joints in both courses.

The tile sill is made 25 or more wider each side than the window opening, the tiles being notched around the angle of the jamb as shown in Fig. 92.

The purpose of this is purely decorative as it is generally considered that a sill looks more attractive if it is wider than the window opening.

The necessary slope or weathering of the sill depends upon the porosity of the tiles used. Hand-made clay tiles fairly readily absorb water and should be laid at a slope of 30 degrees or more to the horizontal. Machine pressed clay tiles and good quality concrete tiles are denser and have been successfully laid at slopes of as little as fifteen degrees to the horizontal. In exposed positions where sills are liable to be heavily saturated by rain followed by heavy frost, a tile sill should be steeply weathered or some other denser material, which is less likely to be damaged by frost, should be used. The position of a window frame in the width of the reveal is partly determined by the type of sill used. For example, plain tiles are 267 long and if whole tiles are used for sills which project 40 then the window frame must be some 200 back from the external face of the wall. If the wall is only 1B. thick this is obviously not practicable and tiles cut down to say 150 lengths or under eaves tiles (see Vol. 1) have to be used.

As an alternative to laying tiles with their length at right angles to the window they are sometimes laid with their length parallel to it as illustrated in Fig. 93. In this way full tiles can be used for sills in walls only 1B. thick and the face of the window frame is some 100 in from the wall face. Creasing tiles are generally used for this type of sill as they have no nibs on them and they are flat. Creasing tiles are 267 long by 165 wide and are made from the same clay as machine made plain clay roofing tiles. A roofing or creasing tile sill is one of the cheapest in use today. Tile sills are usually bedded below windows after walls have been built so that they are not damaged whilst the walls are being built.

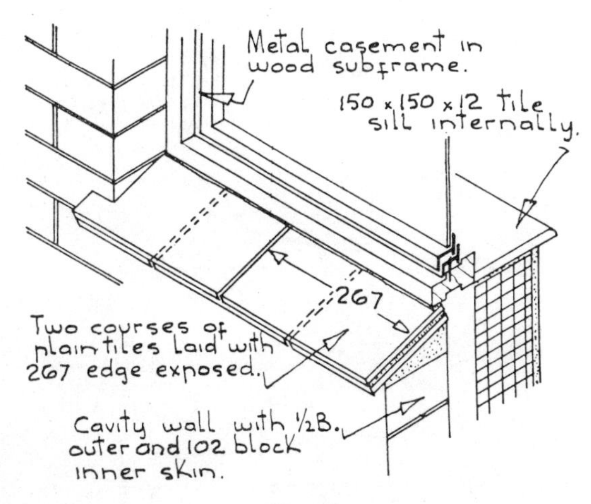


Fig. 93

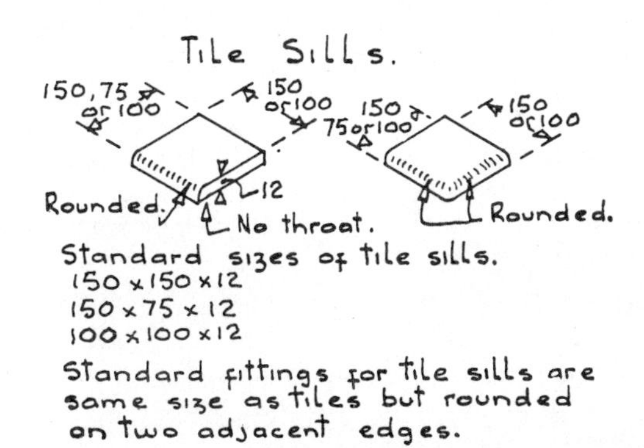


Fig. 95

(2) **Quarry sills—Tile sills.** A range of quarry sills and tile sills is manufactured for use in external or internal sills. The quarry sills and tile sills are made from the same types of clay as floor quarries and plain colour floor tiles respectively. The moulded sills are burned and after burning have the same characteristics as floor quarries or plain colour floor tiles. One edge of these sills is rounded and special return nosing tiles are made, as illustrated in Figs. 94 and 95.

Quarry sills and tile sills have been principally used for internal sills as they form an attractive level surface which is very easily cleaned. They are equally effective as an external sill because they are dense and not liable to damage by frost. When used as an external sill they are laid with a slight slope (weathered) to throw water out. The tiles should be thoroughly soaked overnight by immersion in a bucket or tank of water and then bedded on cement and sand (mix 1:3) and levelled in the same way that floor tiles are laid. The joints between the tiles are filled with neat cement grout. The words cement grout describe a mix of cement powder and water which is washed into the joints and then finished level with the surface of the tiles. Fig. 96 is an illustration of an external quarry sill. Internal quarry and tile sills are usually laid so that they project about one inch beyond the plaster finish, because it is difficult to achieve a neat flush finish between the edges of the tiles and the plaster, on account of slight variations in the shape of the tiles. Quarry and tile sills are usually bedded below windows after walls have been built.

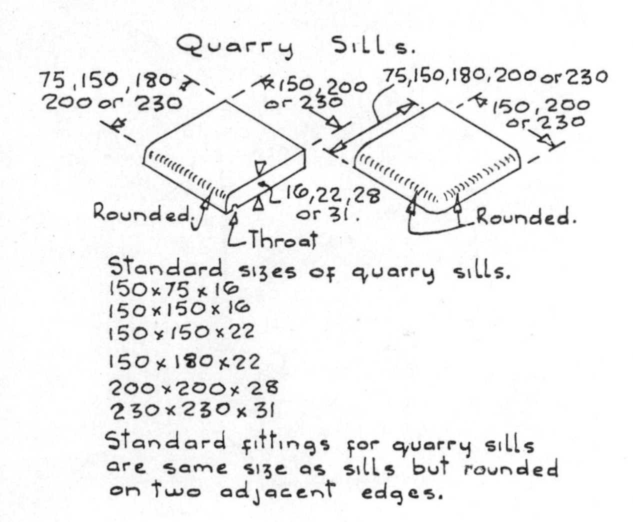


Fig. 94

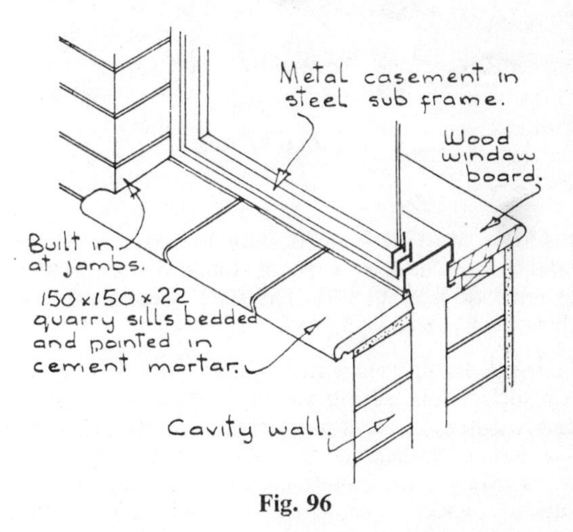


Fig. 96

(3) **Brick-on-edge sill.** Sound, hard burned facing or engineering bricks are laid weathered, on edge in cement mortar as an external sill. The bricks are laid on edge for the same reason that bricks are laid on edge in a brick and tile creasing capping (Vol. 1.). The bricks are laid so that they project about 25 or more beyond

the external face of the wall and so that the other end of each brick is bedded under the window frame as shown in Fig. 97. Providing sound, hard bricks are used and they are laid sufficiently weathered so that they do not become heavily saturated in winter, they form an effective external sill. The disadvantage of this type of sill is that it has a somewhat lumpy unattractive appearance.

It will be seen from Fig. 97 that a D.P.C. is bedded below the brick sill. This will prevent any water soaking through the bricks to the wall below and also assists in keeping the cavity free from mortar droppings as the sill is laid.

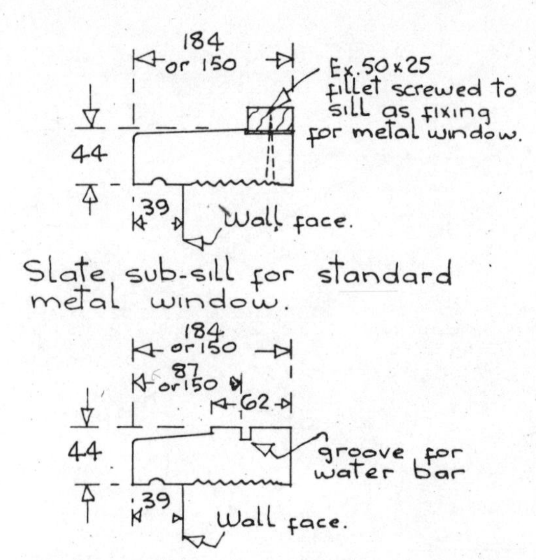


Fig. 98

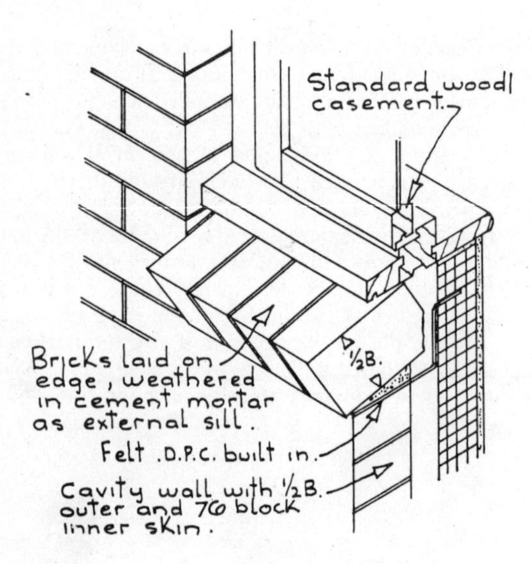


Fig. 97

(4) **Slate sill.** Fine grained, hard Welsh or Westmoreland slate is cut in a variety of standard sizes for use as external or internal sills. The slate is cut to the sizes illustrated in Fig. 98.

Slate sills are finished with a smooth, slightly weathered top surface, one edge is rounded and the underside or bed roughened so that mortar binds firmly to the slate. The slate is hard and dense and does not readily absorb water. Slate is principally used for external sills. Slate sills are bedded in cement mortar, with their rounded edges projecting 25 or more as illustrated in Fig. 99. A combined external and internal slate sill is also marketed and used as illustrated in Figs. 100 and 101.

The natural grey-blue colour of Welsh slate and the natural green colour of Westmoreland slate is an attractive feature of these materials. A slate sill is more expensive than a similar tile or brick sill.

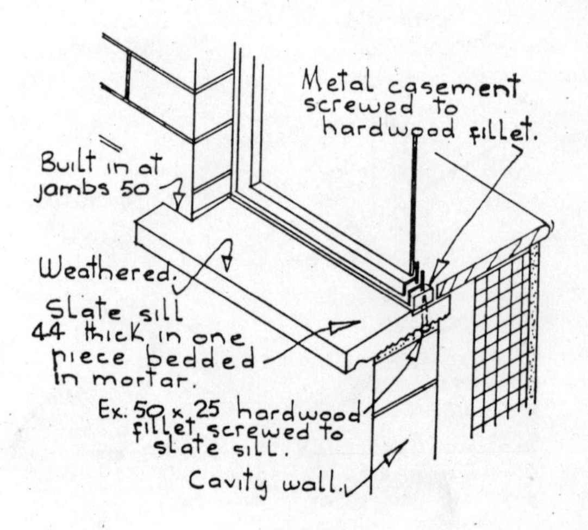


Fig. 99

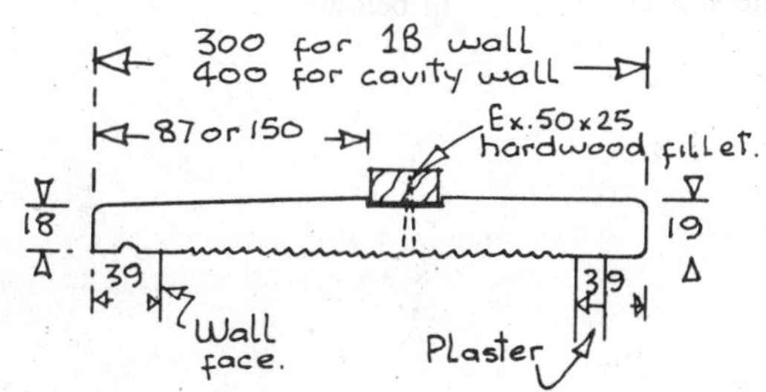


Combined external and internal slate sill for metal window.

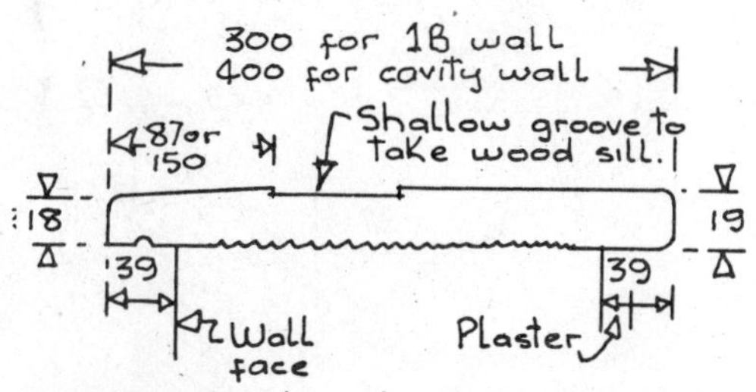


Combined external and internal slate sill for wood window.
Above sills are cut to length to suit width of window, plus 100 to allow for building in at jambs.
Maximum length 2·0

**Fig. 100**

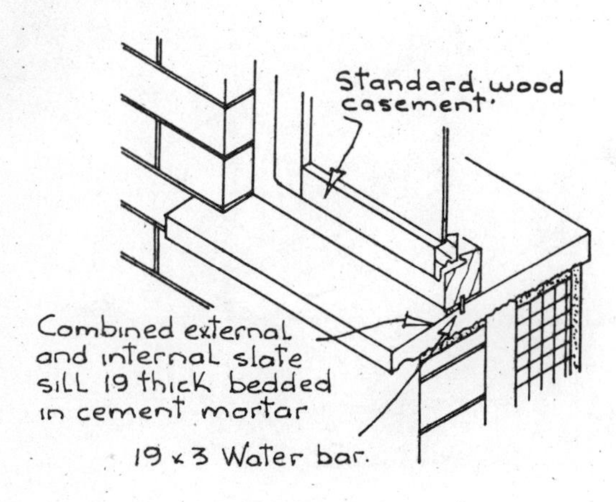


**Fig. 101**

(5) **Stone and cast concrete sills.** A natural stone sill is an expensive form of sill but is often used as the combination of, for example, a stone sill and good quality brickwork is very attractive. A stone sill is cut so that it is sunk, weathered, grooved, throated and stooled at ends as illustrated in Fig. 102. The purpose of the weathering has been explained. A throat is formed on the underside of the projection of the sill for the same reason that one is formed on the underside of coping stones (Vol. 1). The sinking acts as a check against penetration of rain. Both ends of the sill are stooled, and this describes the cutting of a level surface off which the brickwork in the jambs of the opening is built. If the weathering of the sill were continued along the sill to where it is built into jambs, the brickwork built off the sloping surface might slip out of place, hence the stools at the ends. The ends of stone sills are built into the jambs partly to secure the sill in position and also because a sill which is wider than the window opening looks better than one only as wide as the window opening. Because natural stone is expensive, cast stone is often used as a substitute for it.

**Cast stone** is made as described in Chapter 1.

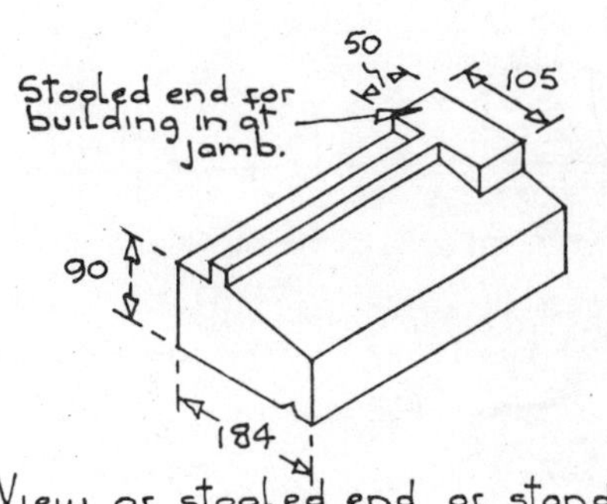


View of stooled end of standard sill for use with metal window.

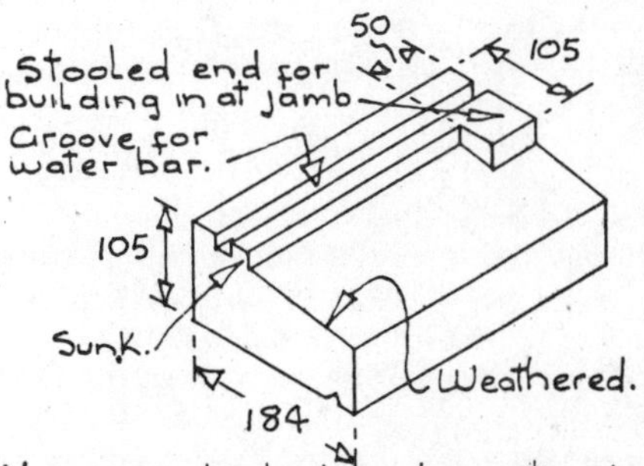


View of stooled end of standard sill for use with wood casement.

**Fig. 102**

A range of natural stone and cast stone or cast concrete sills is cut or cast in standard sections and standard lengths to suit standard metal and wood windows. Fig. 103 is an illustration of the standard sections.

Standard stone & cast concrete sills.

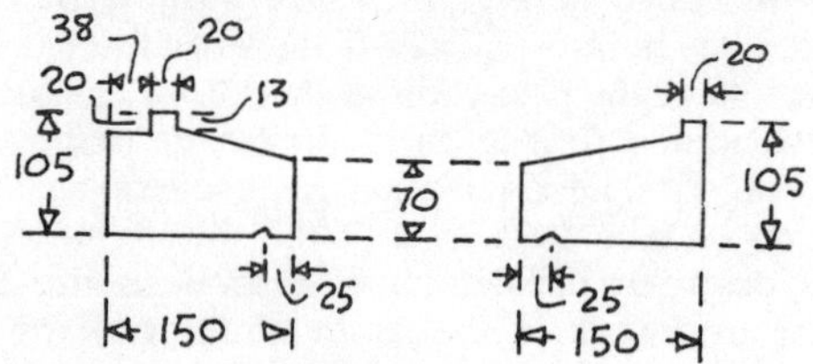


Standard sills for use with metal casement windows.

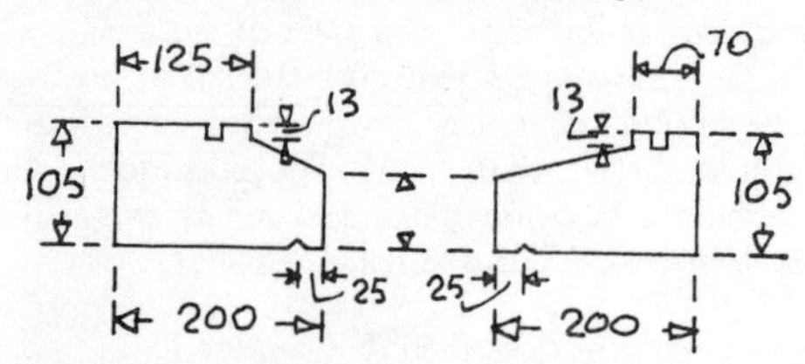


Standard sills for use with wood casement windows.

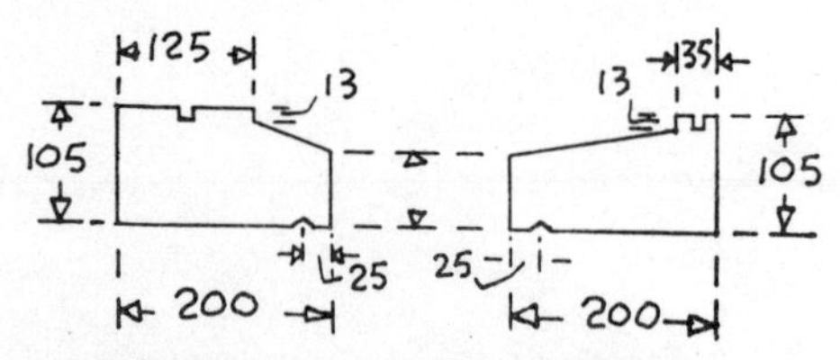


Standard sills for use with vertically sliding sash windows.

**Fig. 103**

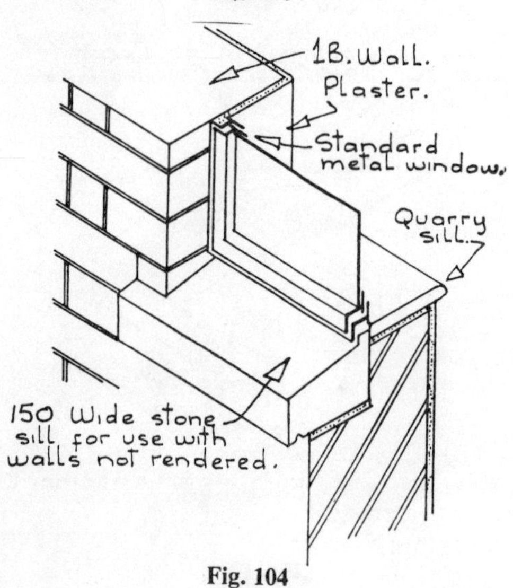


**Fig. 104**

Stone and cast concrete sills are usually built in below windows after brick or block walls have been built. The ends of the sills are built in at jambs and the sill is bedded in mortar.

Figs. 104, 105 and 106 illustrate the use of these sills which are designed for use with either solid or cavity walls with the window set an exact distance in from the wall face.

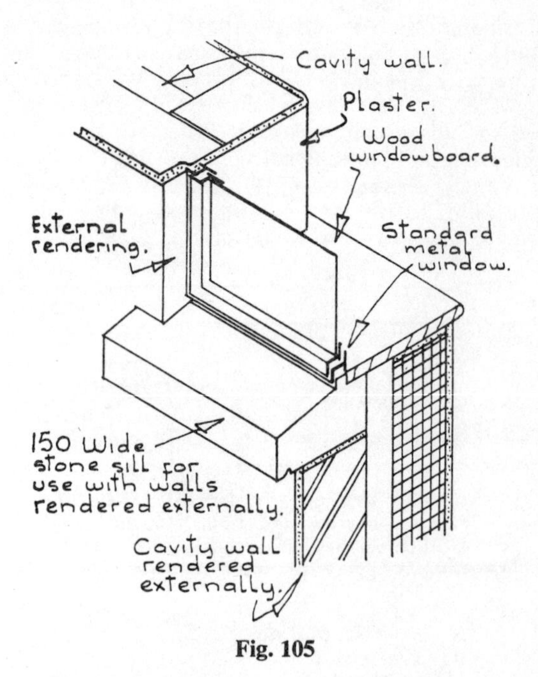


**Fig. 105**

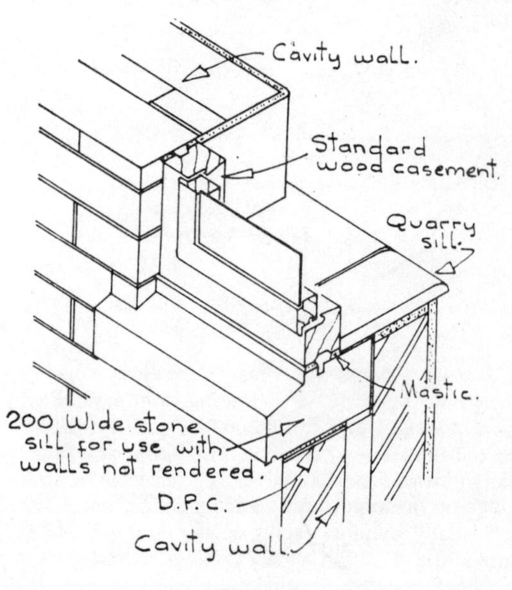


**Fig. 106**

(6) **Clayware sills.** The manufacturers of clay products can supply clay, stoneware, terra cotta or faience sills in standard sections to suit standard metal and wood windows or of special sections to individual designs. These sills, which tend to have a somewhat lumpy appearance and are comparatively expensive, have not as yet been much used. The materials used in the manufacture of these sills are:

(a) *Natural clay:* Any brick earth (clay) which can be moulded, dried and burned to produce a sound, hard facing brick is suitable for use in the manufacture of clayware sills. The clays which have principally been used are those used in the manufacture of facing, engineering and semi-engineering bricks (Vol. 1, Chap. 2). A range of standard section sills can be produced in the sizes shown in Fig. 107 for use with standard metal or standard wood windows.

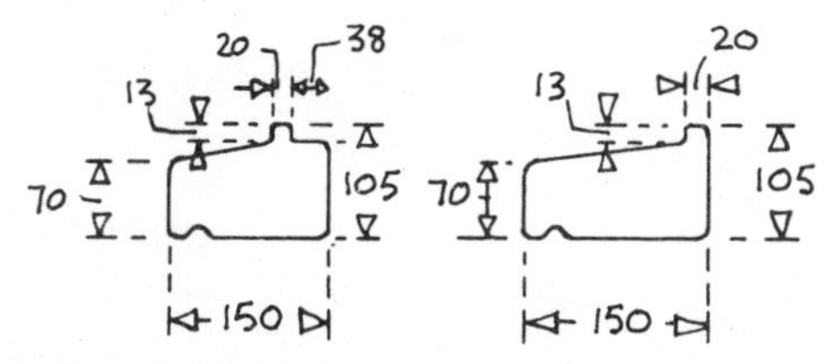


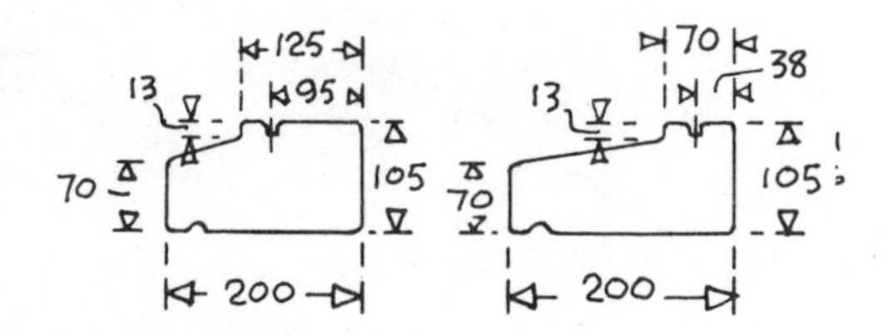


**Fig. 107**

In effect these sills are purpose-moulded sill bricks. Figs. 108 and 108A illustrate their use.

(b) *Stoneware and terra cotta:* Stoneware is the name given to the products of moulding and burning either very sandy clays or a natural mixture of clay and flint after the flints have been crushed. Stoneware does not readily absorb water and does not generally shrink on drying and burning to the same extent as natural clays. It is usually a light biscuit colour and used for the manufacture of drain pipes, copings, sanitary fittings and sills. Stoneware is principally manufactured in the south western and midland counties of England. As it has neither an attractive colour or texture stoneware is usually finished with a glazed surface which is fired on to the material during burning.

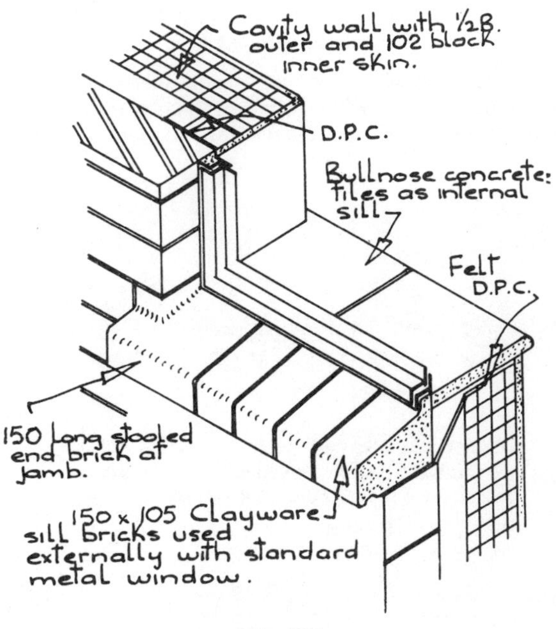


**Fig. 108**

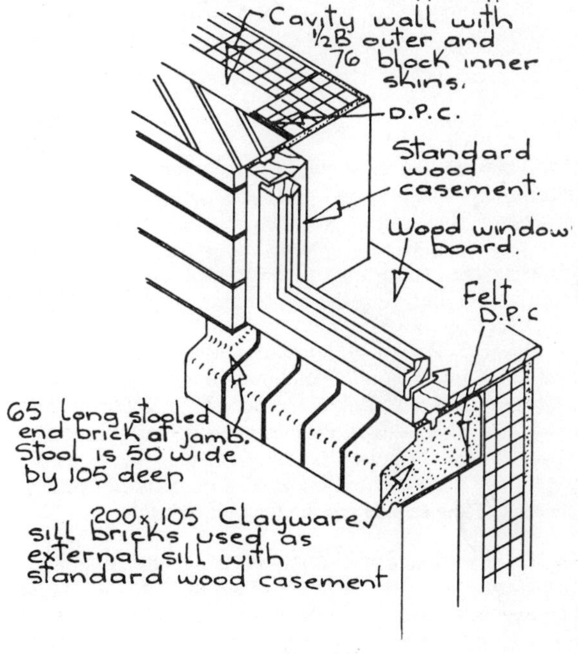


**Fig. 108a**

The words terra-cotta are used to describe the burned product of pure, fine grained clays suitable for pottery making. Terra-cotta is therefore natural clayware. It has a smooth, fine grained finish which is generally semi-vitreous. Its usual colour is brownish red. For sills terra-cotta is either burned with its natural clay finish or may be finished with a glazed surface.

A range of stoneware or terra-cotta sills in standard sizes to suit standard metal or wood windows is manufactured to the sections illustrated in Fig. 109

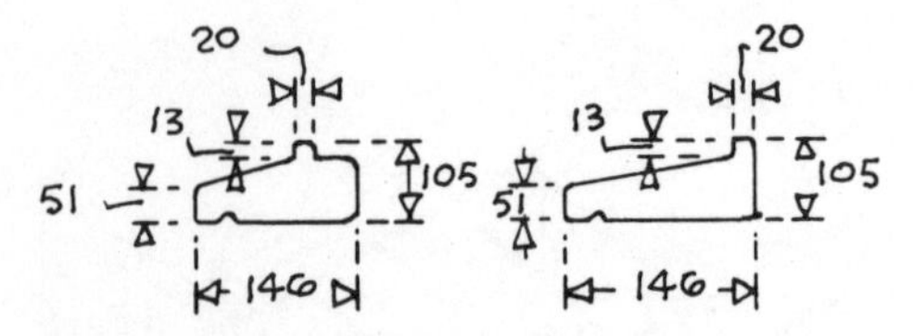


Stoneware and Terra-cotta sills for use with metal windows. Sills are 146, 222 or 298 Long.

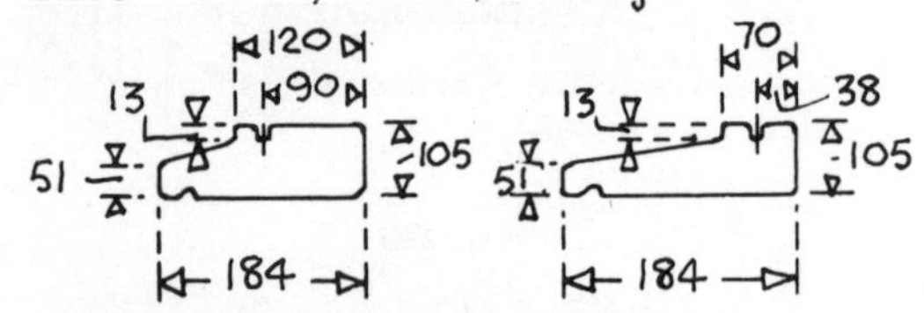


Stoneware and terra-cotta sills for use with wood windows. Sills are 146, 222 or 298 Long.

Fig. 109

It will be seen that these sills are of slighter section than natural clay sills and do not appear so lumpy and unattractive. They can be supplied with a natural finish or glazed to a selected colour. They are dense, non-absorbent and form a hardwearing attractive sill. Figs. 110 and 111 illustrate the application of these sills.

(c) *Faience:* The word faience is used to describe a core or biscuit of moulded and burned fireclay on which a hard non-absorbent glaze has been fired. Fireclay is a sandy clay which does not suffer great drying shrinkage during firing. Because the fireclay does not shrink greatly the glaze burned on to its surface is not likely to crack. Faience sills are in effect glazed tiles.

These sills are made in a variety of sizes to suit standard metal and wood windows as illustrated in Fig. 112. The exposed surfaces of the sills are glazed to selected colours and are used as illustrated in Fig. 112. They form an attractive, hardwearing sill.

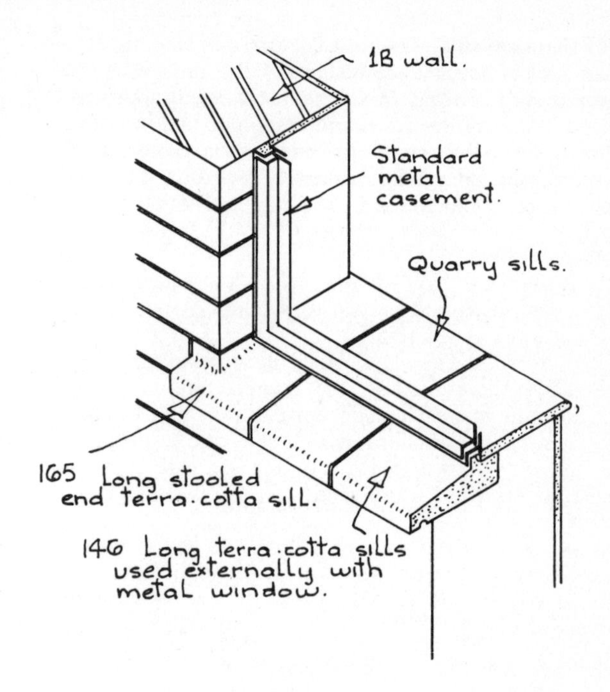


Fig. 110

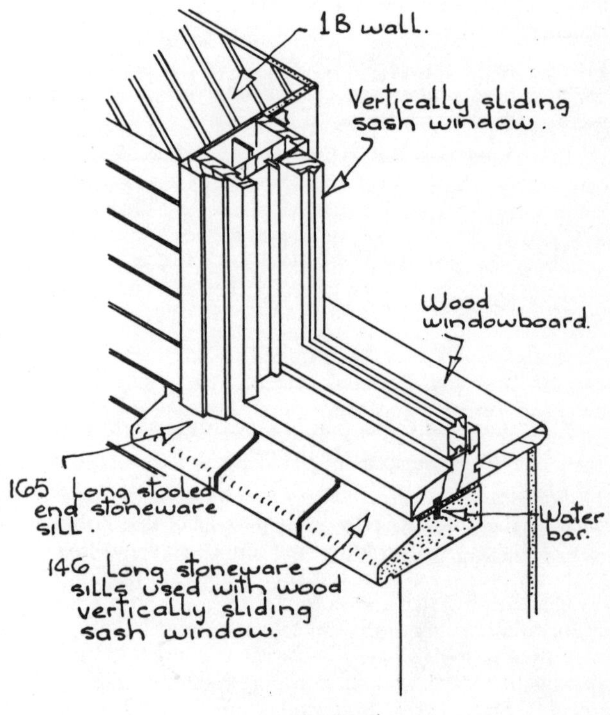


Fig. 111

(7) **Steel sub sill.** The manufacturers of metal windows can supply pressed steel sub sills for use with either standard or non-standard metal windows. In Chapter 2 these sills were described and illustrated. The standard steel sub sills are designed for use with metal windows set an exact distance in from the outside face of walls as illustrated in Chapter 2.

**Internal sills**

(1) **Wood window board.** One usual way of finishing the internal sill of windows is to fix a timber window board. A softwood board is prepared with one edge rounded and it is fixed to grounds, plugs or fixing blocks in the brickwork below the window as illustrated in Fig. 92 and 97. It will be seen that the board projects beyond the face of plaster and has its ends notched around the angle of the reveals. The notching around reveals is done purely for reasons of appearance.

It is not generally possible to drive nails through timber and into brickwork. Most bricks are too hard to be penetrated by nails, hence the use of grounds, plugs or blocks.

Timber grounds consist of lengths of small section sawn softwood size 50 by 25 or 50 by 50. These grounds are either nailed to wood plugs driven into brickwork joints or directly into mortar joints, to provide a level surface to which the window board can be nailed.

Plugs are wedge shaped pieces of timber driven into joints between bricks and to which the window board is nailed.

Fixing blocks are offcuts of lightweight aggregate concrete blocks which are built at intervals into brickwork and into which nails can readily be driven.

(2) **Steel window board.** Metal window manufacturers can supply pressed steel window boards for use with metal windows. The standard sections were illustrated in Chapter 2. The sizes of window board made are designed for use with windows set back a precise distance from the outside face of either solid or cavity walls.

(3) **Concrete sill tiles:** Are manufactured from clean natural sand and Portland cement with a surface of coloured cement and aggregate. The materials are mixed with water and heavily press-moulded. The tiles have one edge rounded. Fig. 113 illustrates a typical sill tile.

The tiles are finished in a variety of colours such as red, brown, green, black and white. Concrete tiles are dense and have hard, easily cleaned surfaces. They are bedded in cement and sand (1:3) and the joints between them filled with a grout of cement. They are used principally as an internal sill in the same way that quarry or tile sills are, with their edges projecting about 25 beyond the face plaster.

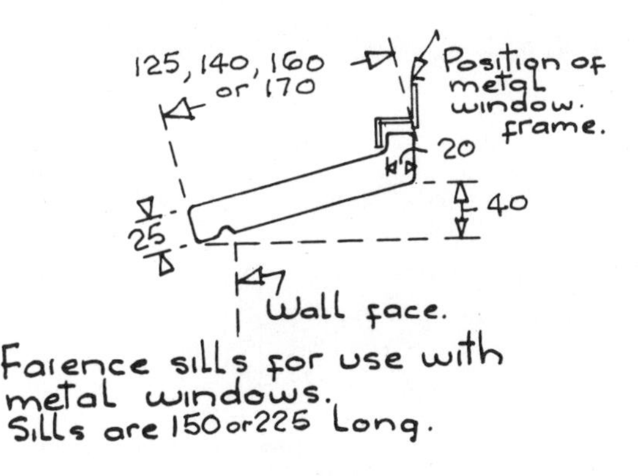


**Fig. 112**

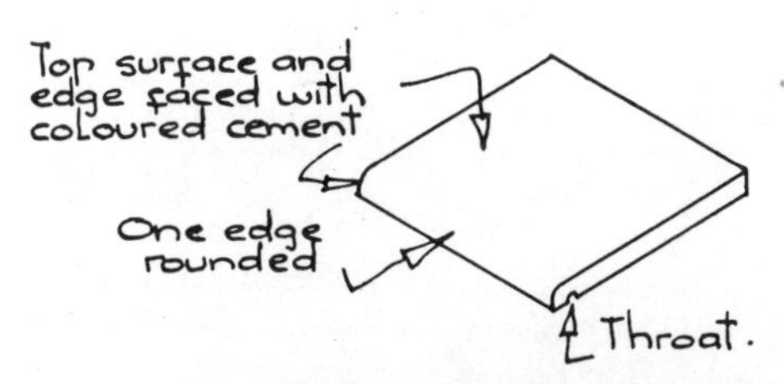


**Fig. 113**

**Reference Books and Publications**

**British Standards:**
No. 402. Clay plain roofing tiles.
No. 473. Concrete plain roofing tiles.
No. 1286. Clay tiles for flooring (and sills).
No. 1422. Steel subframes, sills and window boards.
No. 4374. Sills of clayware, cast concrete, cast stone, slate and natural stone.

# CHAPTER FOUR

# DOORS

## SfB (32) Doors: General

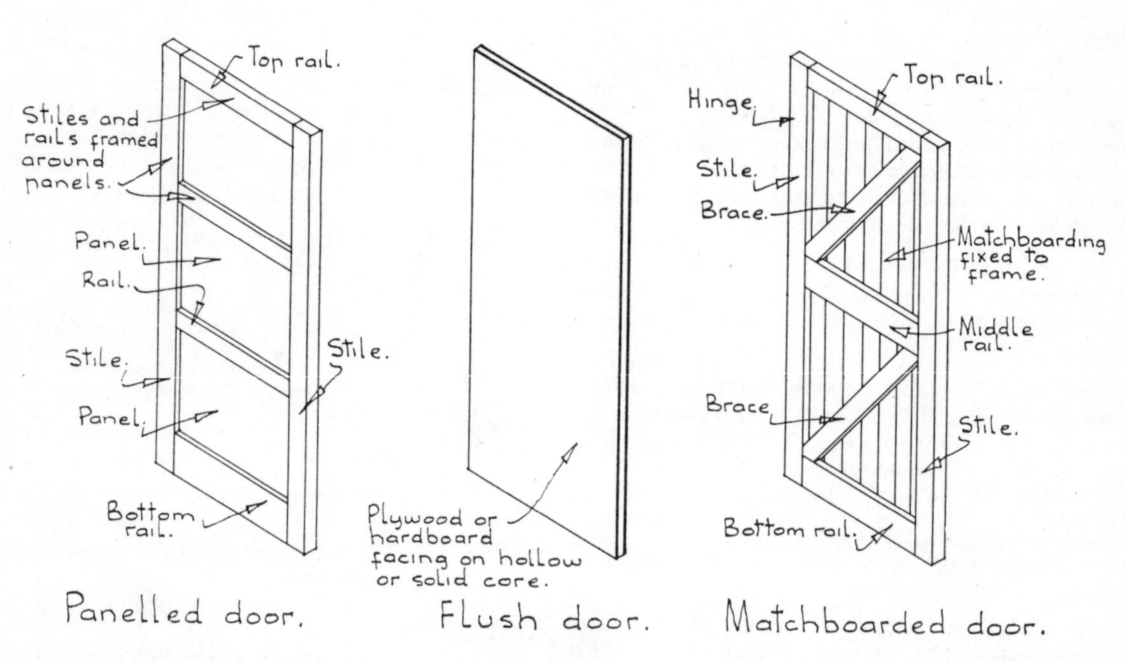


**Fig. 114**

For the change over to metric the British Woodwork Manufacturers Association have proposed to the British Standards Institution a range of Dimensionally Co-ordinated door sets. A door set comprises a machine made door hung in its frame in the factory and delivered to site as a unit. The proposed range of sizes of door sets is designed to fit into dimensionally co-ordinated basic spaces and the proposed metric range of door sizes is designed to this end.

It has been common practice for some years in Scandinavia and the U.S.A., to use door sets. The door, factory hung in its frame, with furniture factory fitted if required and with the frame and door prepared for painting, is delivered to site for building in or fixing in position. The advantage of the use of door sets is a reduction of on site labour and it is claimed that there is an improvement in quality due to factory hanging of the door and fitting of furniture.

The proposed door set sizes overall the frame are 590, 690, 790 and 890, to suit door widths of 526, 626, 726 and 826, and a door set height of 2.09 to suit standard door height of 2.04. The door set widths and heights are for filling basic spaces 10 wider and higher overall than the set sizes thus allowing 5 all round for fixing and jointing.

Doors may be classified as

(1) Panelled doors.
(2) Flush doors, and
(3) Matchboarded doors.

### PANELLED DOORS:

These consist of stiles and rails framed around panels of timber or plywood, as illustrated in Fig. 114. The stiles and rails are cut from timbers of the same thickness and some of the more usual sizes of timber used are: Stiles and top rail 100 by 38 or 100 by 50. Middle rail, 175 by 38 or 175 by 50. Bottom rail, 200 by 38 or 200 by 50. Because the door is hinged on one side to open it tends to sink on the lock stile. The stiles and rails have to be joined to resist the tendency of the door to sink and the two types of joints used are (a) mortice and tenon joint and (b) dowelled joint.

**Mortice and tenon joint:** This is the strongest type of joint used to join timber members framed at right-angles.

Fig. 116 is an illustration of the members of a panelled door before they are put together, cramped, glued and wedged.

**Haunched tenon:** Obviously the tenon cut on the end of the top rail cannot be as deep as the rail if it is to fit to an enclosing mortice, and a tenon about 50 deep is cut. It is possible that the timber of the top rail may twist as it dries out. To prevent the top rail twisting out of upright a small projecting haunch is cut on top of the tenon which fits in a groove in the stile.

Two tenons are cut on the ends of the middle and bottom rails. It would be possible to cut one tenon the depth of the middle rail but the wood either side of the mortice might bow out as illustrated in Fig. 117 and the joint would be weakened. Also a very deep tenon might shrink appreciably and become loose in its mortice. To avoid this, a tenon should not be deeper than five times its thickness, hence the use of two tenons on the ends of middle and bottom rails.

Double tenons are sometimes cut on the ends of the middle rail as illustrated in Fig. 116. The purpose of these double tenons is to provide a space into which a mortice lock can be fitted without damaging the tenons.

It is apparent that the joints between the deep middle and bottom rails and stiles are stronger than that between top rail and stile because of their greater depth of contact with stiles. For strength it would seem logical to make the top rail as deep as the other rails.

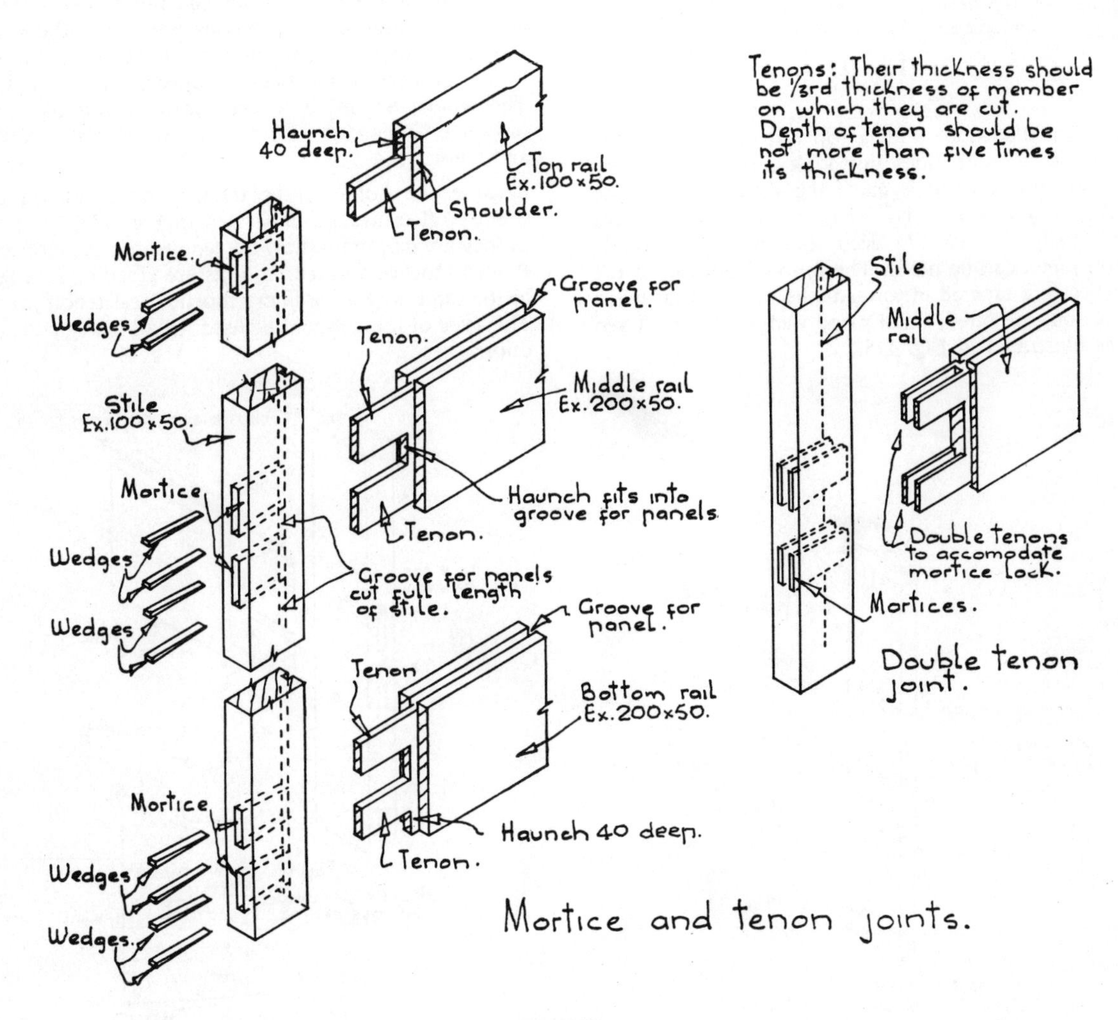


**Fig. 116**

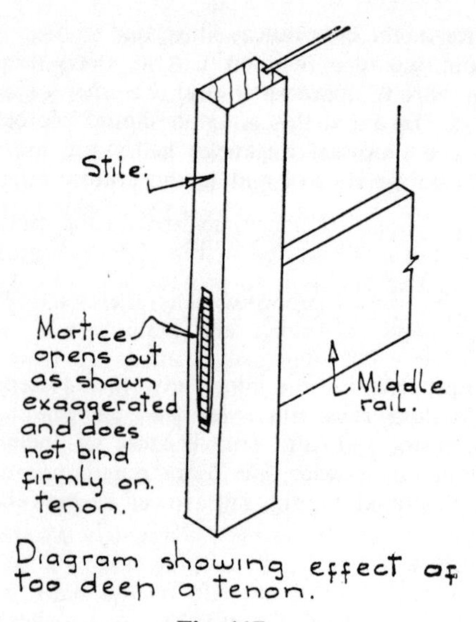


Fig. 117

But a panelled door with a top rail deeper than the width of the stile does not look attractive and by tradition the top rail is made the same depth as the width of the stiles. The middle and bottom rails are usually at least 175 deep so that strong double tenon joints can be made. The top and bottom of the mortices are tapered in towards the rails so that when the tenons are fitted, small wood wedges can be driven in as illustrated in Fig. 118.

**Cramping, glueing and wedgeing:** The wood cramp describes the operation of forcing the tenons tightly into mortices. The members of the door are cramped together with metal cramps which bind the members together until the glue in the joints has hardened. Before the tenons are fitted into the mortices both tenon and mortice are coated with glue. When the members of the door have been cramped together small wood wedges are knocked in the mortices at the top and bottom of each tenon. When the glue has hardened the cramps are released and the projecting ends of tenons and wedges are cut off flush with the edges of stiles.

**Pinned mortice and tenon joints:** If the timber from which a door is made shrinks, the mortice and tenon joints may in time become loose, and the door will lose shape. To prevent this, panelled doors are sometimes put together with pinned mortice and tenon joints. The mortices and tenons are cut in the usual way and holes are cut through the tenons and the timber at sides of mortices, as illustrated in Fig. 119. The tenons are fitted to the mortices and oak pins (dowels), 12 diameter, are driven through both mortice and tenon.

Because the holes in the tenons and stiles are cut slightly off centre, as illustrated in Fig. 119, the pins, as they are driven in, draw the tenons into the mortices. Pinned mortice and tenon joints are glued and wedged in the same way as ordinary mortice and tenon joints. This type of joint should be used for all heavy panelled doors.

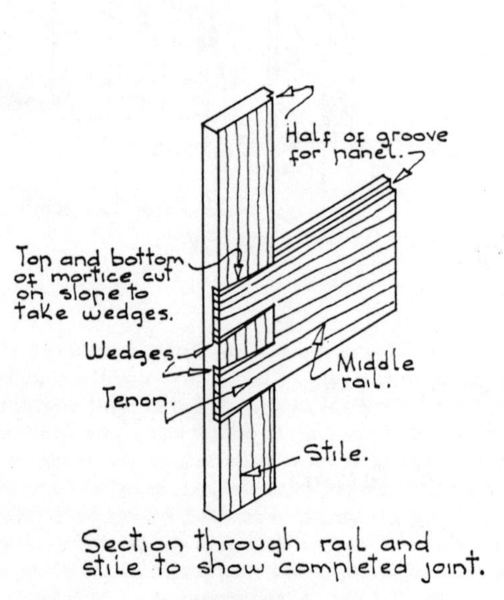


Fig. 118

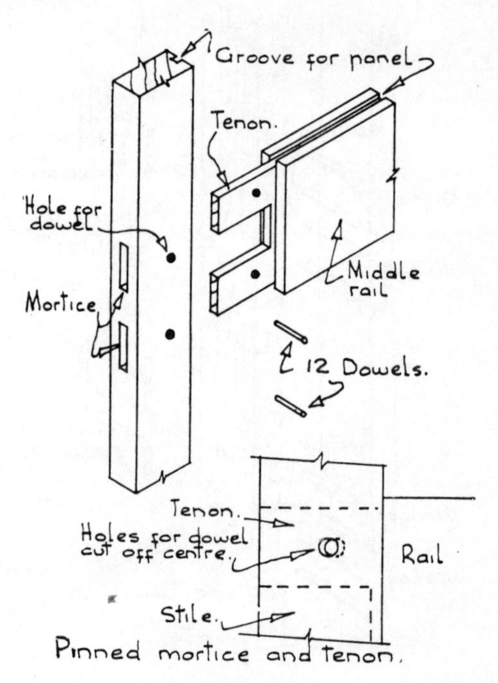


Fig. 119

**Dowelled joints:** The number of operations involved in cutting and assembling mortice and tenon joints by machine is considerably greater than those required for a dowelled joint. To reduce the cost of cutting and jointing the members of panelled doors dowelled joints are commonly used today.

Rails are joined to stiles with round section wood dowels glued and driven into holes cut in the members as shown in Fig. 120. The dowels are of hardwood or the same wood as the members to be joined. They are usually 16 in diameter and 130 long and each dowel is fitted half into stile and half into rail. At least two dowels are used for each joint between top rails and at least three for each joint between middle and bottom rails and stiles. The dowels should be spaced not more than 60 apart, measured between their centres. Two shallow grooves are cut along the length of each dowel so that when it is driven into the holes in stile and rail excess glue and air trapped in the holes can escape.

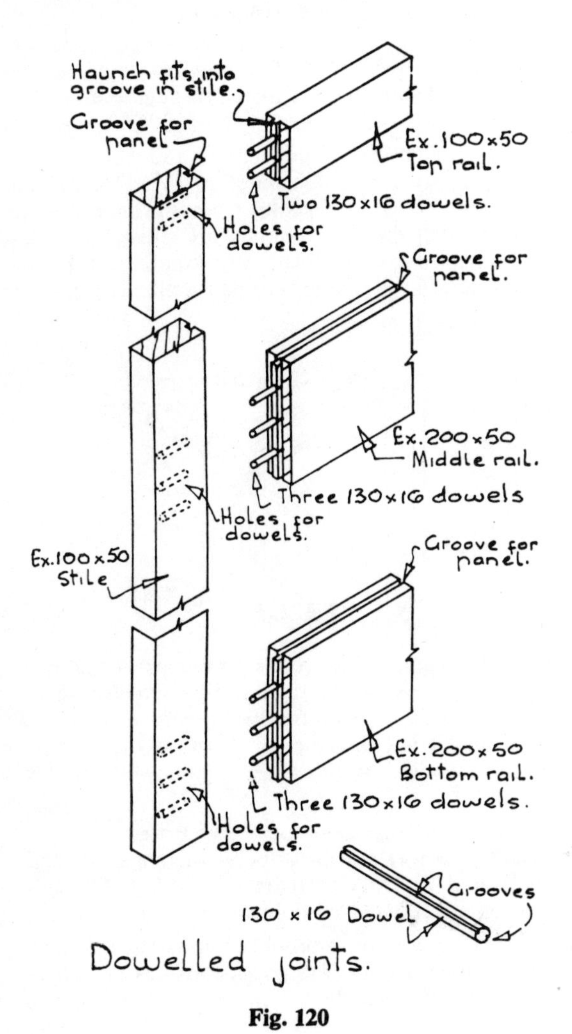


**Fig. 120**

Because of improvements in the binding power of glues this type of joint will as strongly frame the members of panelled doors as a mortice and tenon joint. There is some saving of timber if dowelled joints are used instead of mortice and tenon joints, as the rails can be cut to length to fit between stiles, whereas they have to be cut to at least the overall width of door for mortice and tenon joints.

**Panels:** Before plywood was manufactured, panels were made of timber and plain panels, more than say 250 wide, were usually made up from 12 thick boards, 150 wide, tongued together. The term "tongued" describes the operation of jointing boards by cutting grooves in their edges into which a thin tongue, or feather, of wood is cramped and glued as illustrated in Fig. 121. If the timber from which panels are made has been adequately seasoned, wood panels will not crack or wind and will be perfectly satisfactory during the life of the door. But today plywood is generally used for panels because it is stronger than wood of similar size and not liable to crack or twist.

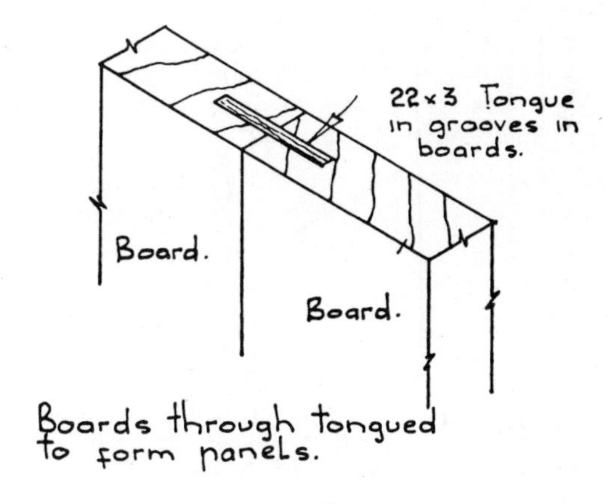


**Fig. 121**

**Plywood:** This is made from three, five, seven or nine plies, or thin layers of wood firmly glued together, so that the long grain of each ply is at right angles to the long grain of the plies to which it is glued, as illustrated in Fig. 122. The most pronounced shrinkage in wood occurs at right-angles to the long grain and any shrinkage of the centre ply is resisted by the outer plies and vice versa. Plywood does not shrink appreciably and because of the opposed long grains of the plies, it does not warp or twist. Three ply wood, 5 or 6·5mm thick, is generally used for door panels.

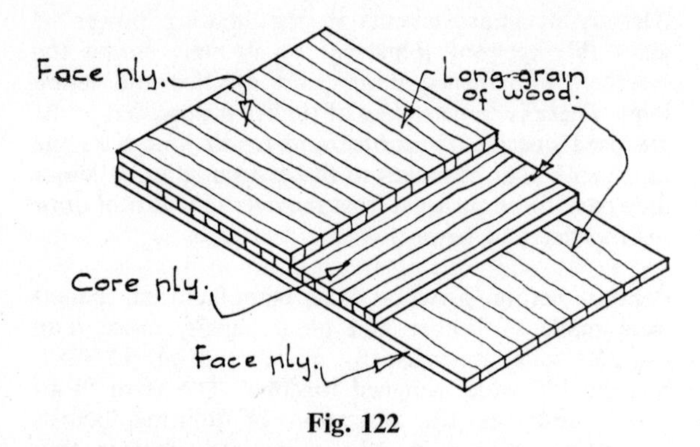


Fig. 122

**Fixing panels:** The usual means of securing panels is to set them in grooves cut in the edges of the stiles and rails as illustrated in Fig. 123. If any shrinkage of the members of the door occurs gaps will not appear around the panels.

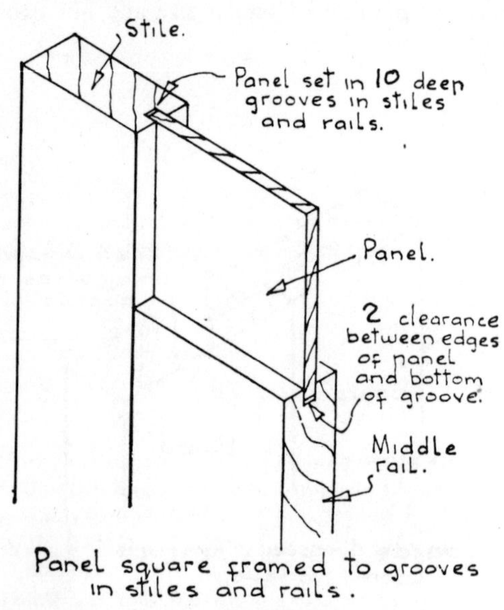


Fig. 123

Another method of fixing panels is to plant (nail) timber beads to the edge of stiles and rails both sides of each panel as illustrated in Fig. 124. The disadvantage of this method is that the beads may shrink and in time the panels may rattle and also ugly cracks may appear around the beads due to shrinkage.

A door with plain (flat) panels set in grooves in stiles and rails which have square edges, as illustrated in Fig. 123, looks unfinished and it is usual to cut some form of moulding on the edges of stiles and rails around each panel. A moulding cut on the edges of stiles and rail is described as a stuck moulding. A stuck moulding around panels is illustrated in Fig. 125.

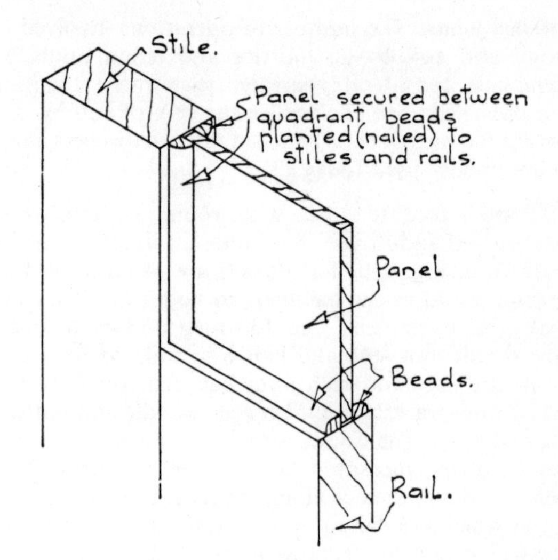


Fig. 124

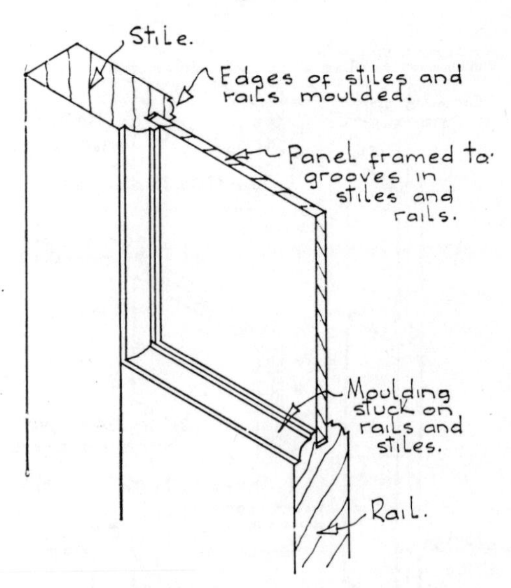


Fig. 125

The traditional panelled door was constructed with four or six panels as illustrated in Fig. 126. The centre vertical member separating the panels is called a muntin and it is either the same size as the stiles or 25 less in width. Muntins are usually stub-tenoned to rails.

But fashions change and today the British Standard (B.S. 459, Part 1) panelled door is arranged without muntins and with one or more panels. Fig. 127 illustrates some B.S. panelled doors.

Because they are mass-produced in standard sizes, British Standard doors are considerably cheaper than similar non-standard doors, and are very extensively used.

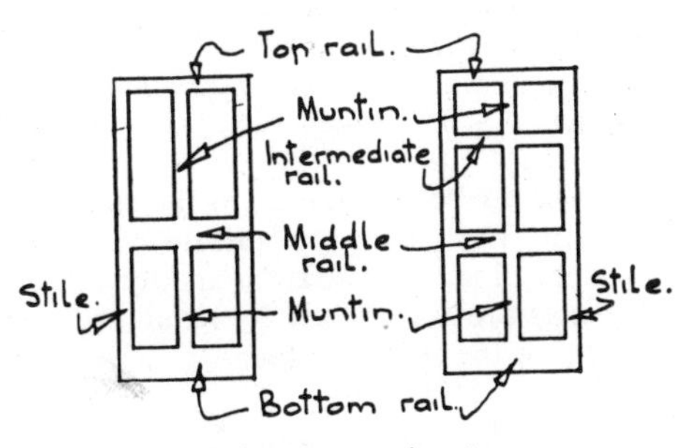


Fig. 126

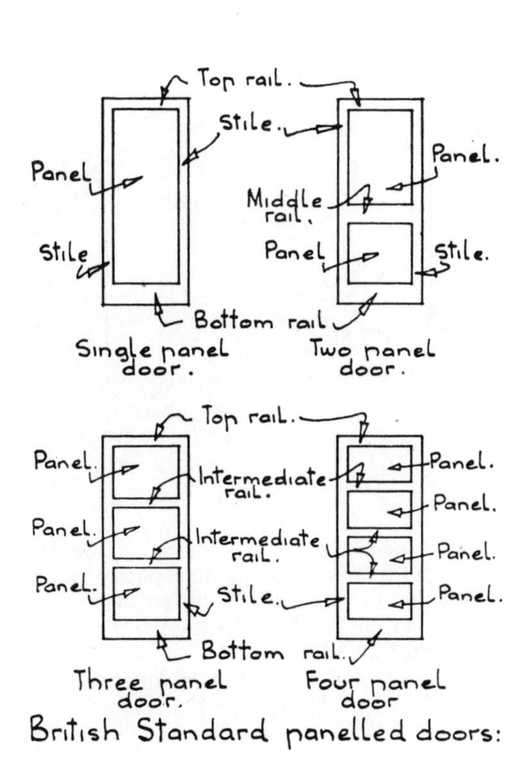


Fig. 127

The top, bottom and middle rails are through-tenoned to mortices in stiles or jointed with at least two dowels to stiles. Intermediate rails are either stub-tenoned or jointed to stiles with at least one dowel.

**Stub-tenon** describes a tenon fitted to a mortice not cut right through a member. The tenon is glued and cramped into the mortice. Fig. 128 illustrates a typical stub-tenon.

The panels of the B.S. doors may be timber, plywood or hardboard set in grooves in the stiles and rails. There should be a clearance of 2 around each panel when it is set in the grooves, so that shrinkage of the stiles and rails around the panel can occur without buckling the panel. A moulding is cut on stiles and rails around each panel.

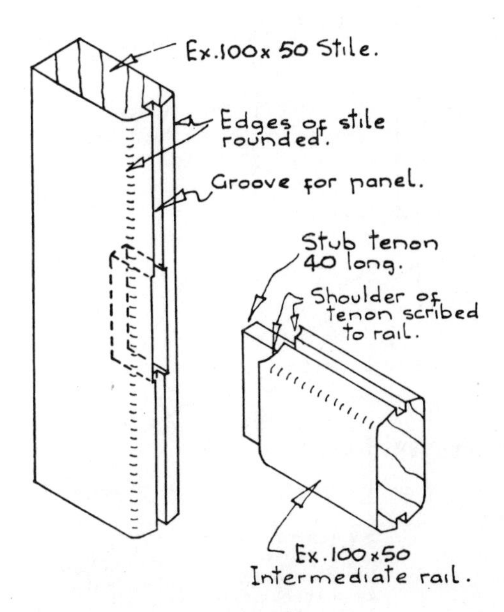


Fig. 128

**Doors with raised panels:** Large doors, such as entrance doors to banks and public buildings and internal doors to council chambers and boardrooms, are often made more imposing than the normal panelled door by cutting the panels to be other than absolutely flat. The panels are usually cut so that they are thicker at their centre than at their edges and are described as raised panels. The usual types of raised panels are:

*Bevel raised.* Each panel is cut with four bevel faces on each side, the bevel faces having the same slope so that they rise to a central ridge, if the panel is rectangular, or a point if it is square, as illustrated in Fig. 129.

*Bevel raised and fielded.* Instead of the bevel rising to the centre of the panel it rises to a flat surface in the centre of the panel, called the field. At the point where the bevel meets the field a slight sinking is cut to emphasise the change of plane. Again the four bevel planes on a panel have similar slope so that the field is rectangular or square according to the shape of the panel. At the field the panel is either as thick as the stiles and rails of the door, or slightly thinner. This is the most commonly used type of raised panel and the proportion of the area of fielded surface to the whole panel is a matter of taste. Fig. 130 is an illustration of this type of raised panel.

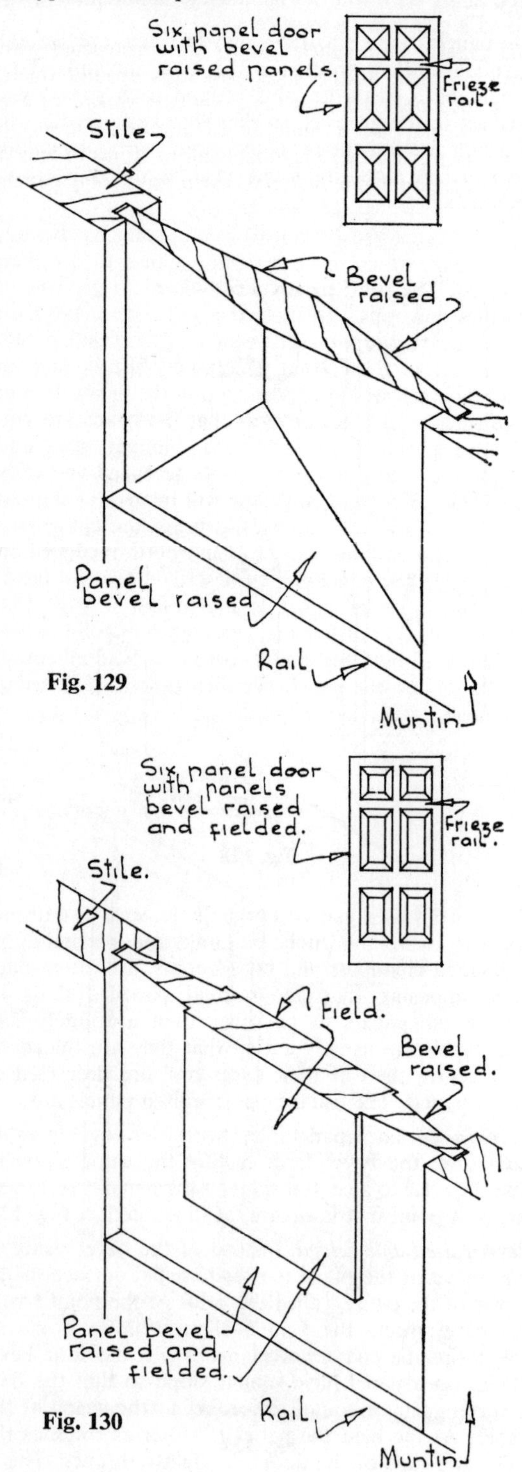


Fig. 129

Fig. 130

*Raised and fielded panel.* Instead of the panel rising from its edges in the form of bevelled surfaces, it rises, that is increases in thickness, at right-angles to the face of the panel to a flat fielded surface, as illustrated in Fig. 131.

Raised panels are set in grooves in the stiles and rails around them. The panels are not always raised both sides and sometimes one side is flat and the other raised. Raised panels are usually cut by hand and they add considerably to the cost of a door.

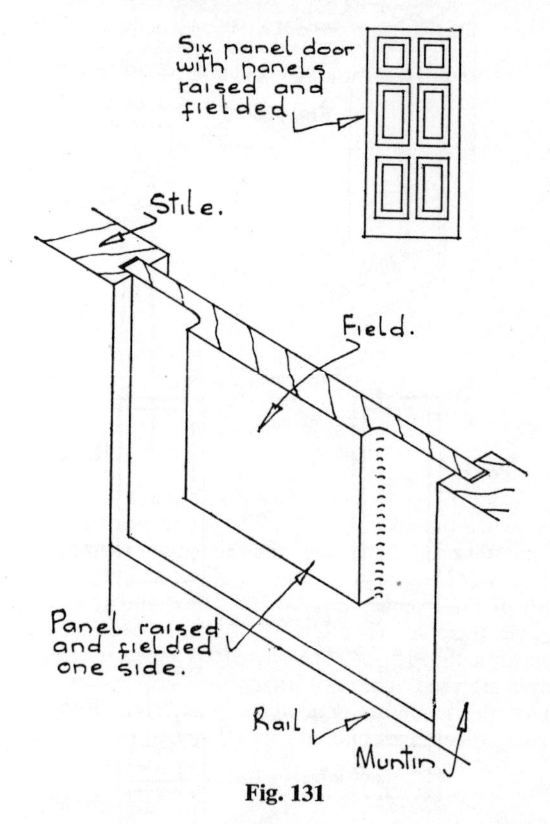


Fig. 131

**Bolection moulding:** A timber bolection moulding is planted around the panels of a door purely for decoration. A bolection moulding is cut so that when it is fixed it covers the edges of the stiles and rails around panels as illustrated in Fig. 132.

It will be seen that the bolection moulding projects beyond the face of the door and as it emphasises and decorates panels it is generally used around raised panels.

**Hardwood panelled doors:** Entrance doors and doors to the main rooms in large buildings are often constructed of some hardwood with a good colour and decorative grain, such as mahogany. These doors are polished or french polished so that the colour and grain of the wood can be seen through the polish film. It would spoil the appearance of such doors if the ends of the tenons cut on the rails appeared on the edges of the

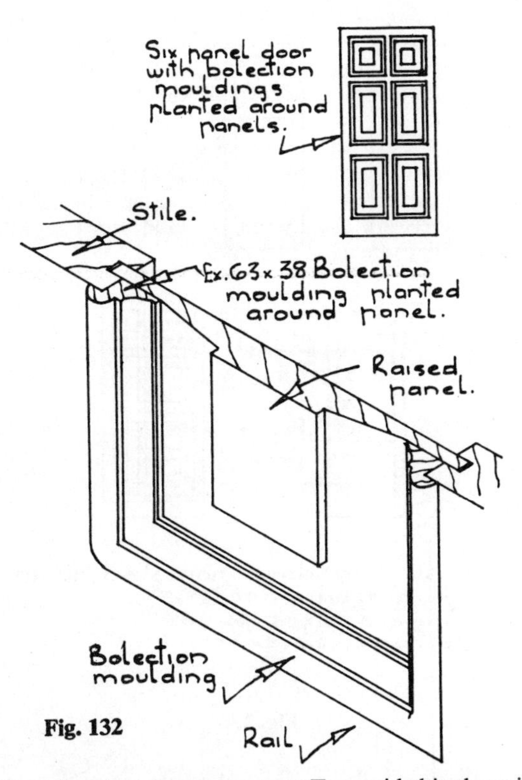

Fig. 132

stiles, when the door was open. To avoid this the rails are stub-tenoned to the stiles and the tenons are secured with foxtail wedges. A stub-tenon is not as long as the width of the member into which it fits, and it is not possible therefore to wedge it top and bottom as a tenon in a through-mortice is wedged. Instead foxtail wedges are used. They are driven into saw cuts in the end of the tenon so that when it is driven into its mortice, the wedges bite into it and spread it so that it binds in the mortice. This is illustrated in Fig. 133. This joint has to be very accurately cut because if the wedges are too long they may split the rail and if too short they will not spread the tenon sufficiently to bind it firmly in its mortice.

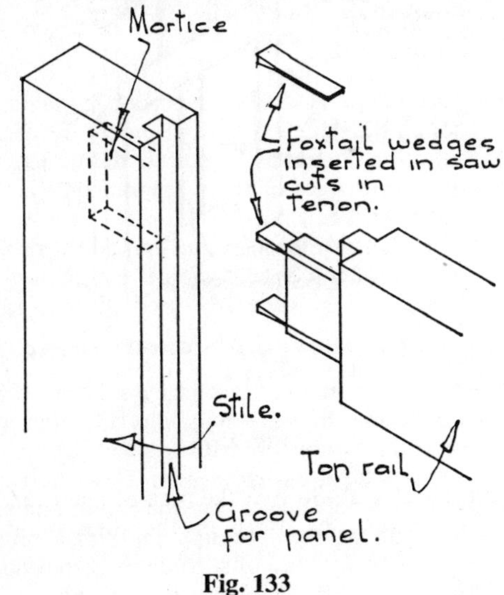

Fig. 133

**Solid panels—flush panels:** Doors are constructed with panels as thick as the stiles and rails around them, where the doors act as a barrier to the spread of fire from one part of a building to another. Doors to electricity transformer chambers and to stores in which there are inflammable materials often have solid panels also.

These doors are usually constructed of some hardwood, such as oak, which has a better resistance to fire than softwood. The solid panels are tongued to grooves in the stiles and rails and are either cut with a bead on their vertical edges, or with a bead all round each panel. The timber of these solid panels may shrink and an unsightly crack would appear around the panel. It is to make such cracks less obvious that the beads are cut.

Timber shrinks more across than along its grain and because the long grain of panels is arranged vertically the shrinkage at sides of panels will be greater than at top and bottom. For this reason beads are cut on the vertical edges of panels and top and bottom edges butt to the rails. The panel is then described as solid bead-butt. This type of panel is illustrated in Fig. 134. Some people do not like the appearance of solid panels finished bead-butt and prefer to have a bead all round each panel and the panels are then described as being bead flush.

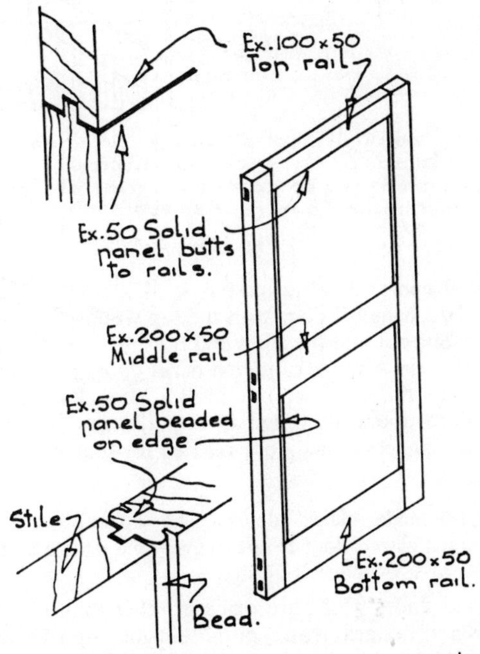

Fig. 134

It is not practicable to cut a neat bead on the end grain of the wood of panels, so the beads are either cut on the edges of the rails or a bead is planted in a rebate as illustrated in Fig. 135. A planted bead is not generally used for external doors as water may get behind it and the beading swell and come out. The rails of doors with solid panels are tenoned to mortice in the stiles as explained previously for panelled doors.

Door panels are not always finished exactly the same both sides and combinations of plain, solid and raised panels are sometimes used. For example, panels are sometimes finished as solid bead-butt outside and plain recessed inside to economise in timber.

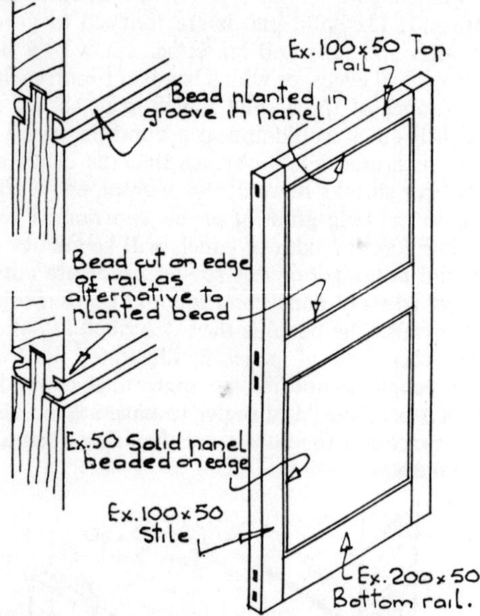


Fig. 135

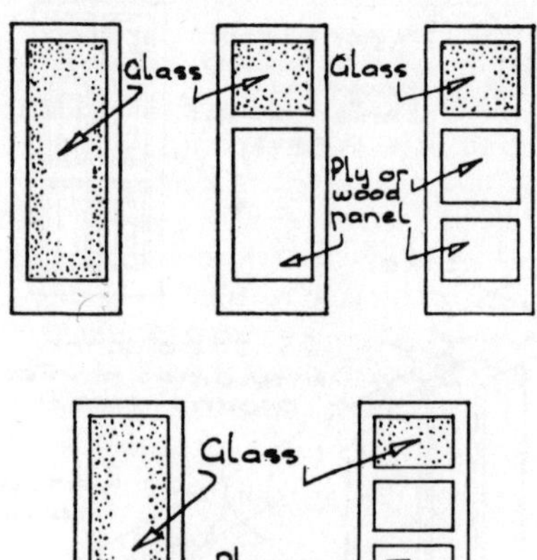


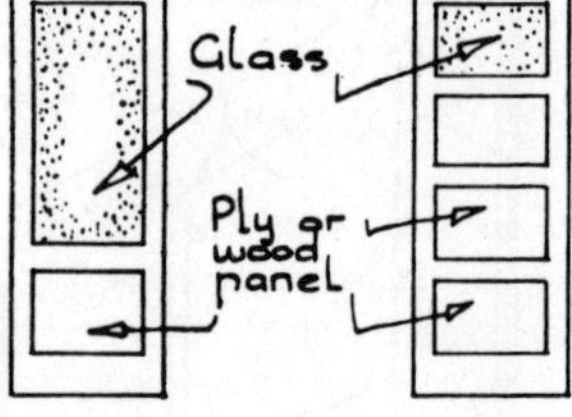


British Standard panelled doors with glazed panels. Standard sizes as for B.S. Panelled doors.

Fig. 136

**Glazed doors:** The cheapest type of glazed door is a standard panelled door with the top wood or plywood panel left out and the stiles and rails around it rebated for glazing. A range of glazed panel doors is manufactured to the sizes set out in B.S. 459, Part 1. The standard doors illustrated in Fig. 136 are made with the top panel of each door rebated for (putty or bead) glazing.

**Purpose-made glazed doors:** The appearance of a standard glazed door is not to everyone's taste and the range of sizes made is limited.

Glazed doors which are made in other than standard sizes and designs, that is purpose-made, are often constructed with diminishing stiles. Fig. 137 is an illustration of this type of door. It will be seen that the lower panels are of timber with a large glazed panel

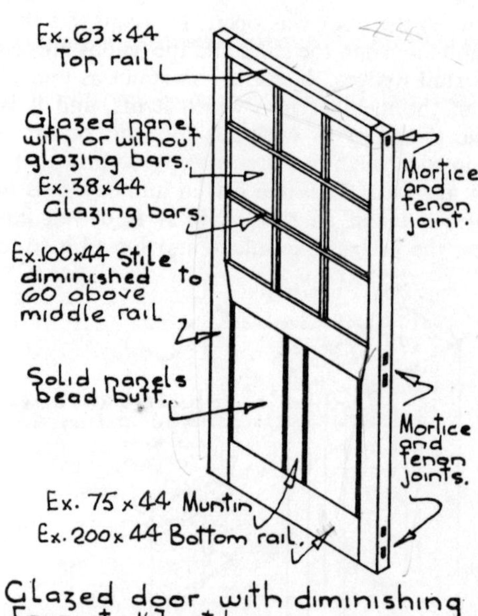


Fig. 137

above the lock rail. So that the area of glass can be as large as possible, the width of the stiles reduces or diminishes from 95 to 60 at the lock rail. The stiles are shaped like the stock of a sporting gun and are 'sometimes described as gunstock stiles. The rails of this door are usually joined to the stiles with glued and wedged mortice and tenon joints as previously described, the only difference being that the tenons on the middle rail have to be cut with splay shoulders to fit the splay of the stiles where they diminish in width. The glass can be fixed either with putty or beads. If the door is made of hardwood which is to be varnished or polished then beads will be used to secure the glass as previously described. Some say that a door should be glazed with beads, as putty is liable to crack and fall out due to the vibration caused by constant closing of the door. It is cheaper to fix glass with putty than with beads.

The upper part of this door can be glazed with one square of glass, or several squares in glazing bars. Glazing bars add slightly to the cost of a glazed door, they somewhat obstruct light and make it more difficult to clean the glass, but many people prefer small squares of glass to one big square. The principal reason for this preference is that glazed doors with glazing bars are associated with the so-called Georgian style of building. During the early part of this period of building large sheets of glass were not manufactured and glazing bars had to be used. The craftsmen of that period produced such happy proportions in their work that many choose to reproduce them today.

Glazing bars are cut from timber as thick as the members of the door and 25 or 38 wide. The bars are rebated and moulded to the same section as the stiles and rails and are through-tenoned to stiles, stub-tenoned to rails and stub-tenoned together at inter-sections as illustrated in Fig. 138.

**Fully glazed door or French casement:** Many of the windows of French houses are arranged as casements with the sill of the window at or just above floor level. Doors, in this country, which are glazed between the bottom rail and top rail are often known as french casements or french doors. Fig. 139 is an illustration of a typical french casement with glazing bars. The stiles and top rail are usually cut from 75 by 50 or 100 by 50, and the bottom rail out of 175 by 50 or 200 by 50 timbers rebated for glass and moulded inside. Rails are jointed to stiles with glued and wedged mortice and tenon joints as previously described. Horizontal glazing bars are through-tenoned to stiles and vertical ones stub-tenoned to rails.

Glazing bars are cut from timber as thick as the main members of a door and some 25 or 38 wide. The bars are rebated and moulded to the same section as the stiles and rails and are through-tenoned to stiles. stub-tenoned to rails and half-lapped or tenoned at ntersections.

Standard casement doors are made to the sizes illustrated in Fig. 140. This is the cheapest type of french casement available and can be supplied with or without glazing bars.

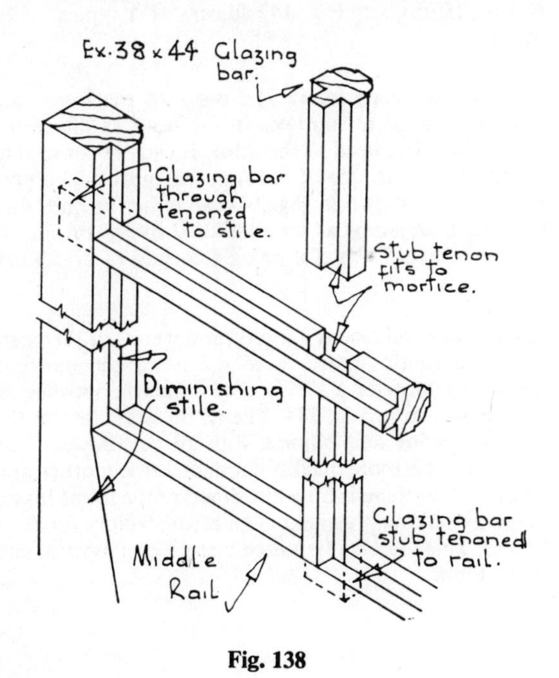


Fig. 138

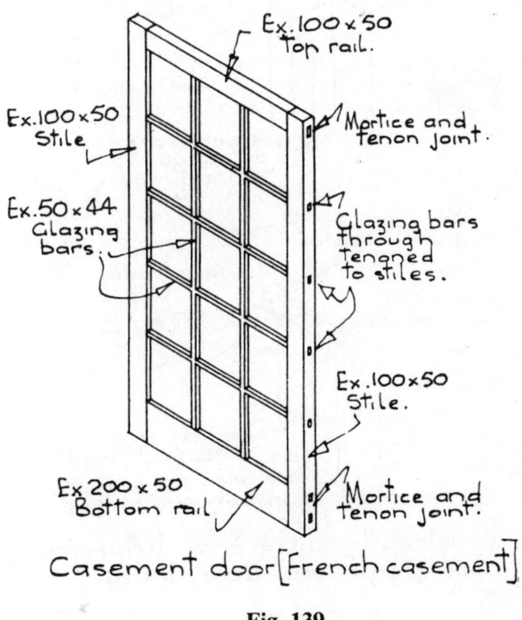


Fig. 139

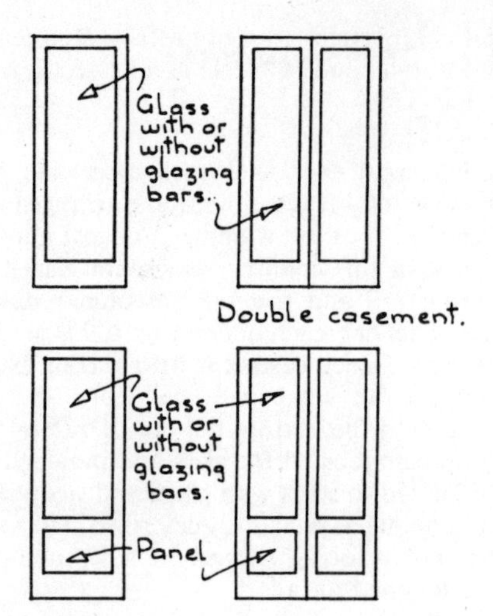


Fig. 140

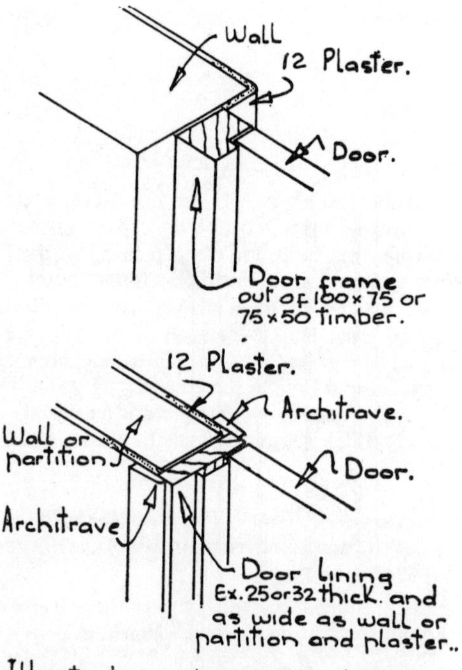


Fig. 141

**Door frames and linings.**

A door frame is made from rectangular section timbers in which a rebate is cut or to which a stop is nailed into which the door closes. The members of the door frame are usually about twice as wide as the door is thick.

A door lining is made from timber boards 25 or 32 thick and as wide as the reveal of the opening in which the door is hung. It is called a lining because it lines the reveal of the door opening. Fig. 141 illustrates the difference between a door frame and a door lining.

Door frames, which are made from more substantial timbers than linings, are used for most external doors, for larger internal doors and for doors in thin non-load bearing partitions. Door linings are used for internal doors in brick partitions and load bearing partitions. A door frame which is as wide as the reveal of the opening in which it is fixed is in effect a lining, as it lines the reveal. It is often difficult to distinguish between frames and linings. The essential difference is that a lining is made of thin boards and a frame of larger section timbers.

A door frame consists of three or four timber members rebated 12 deep for the door or with planted stops, and joined with mortice and tenon or slot mortice and tenon joints. Door frames are used for external and internal doors and are built into brick or block walls and secured with "L" shaped building-in lugs. The lugs are of galvanised or wrought iron, are 150 by 75 by 38 and are screwed to back of frame through the 75 arm, with the 150 arm built into horizontal joints. Doors are hinged on and close into the frame. Frames for external doors are cut from 100 by 75 or 100 by 50 timber. Fig. 142 illustrates a typical door frame.

**Mortice and tenon joint:** The posts of the frame are usually tenoned to mortices in the head as illustrated in Fig. 143. The head of the frame is usually cut so that it projects either side of the frame as horns. The projecting horns shown in Fig. 143 can either be built into brickwork at jambs as an additional fixing, or cut off on site to avoid wasteful or ugly cutting of brickwork around the horns.

**Slot mortice and tenon:** Tenons on the ends of the posts are fitted to slot mortices in the head, the mortices taking the form of a slot instead of a true mortice or hole as shown in Fig. 144. The tenon is secured in the slot by glueing and pinning with an oak dowel. This type of joint is more quickly cut than a true mortice and tenon and, as there is little likelihood of the frame losing shape by sinking, this joint is quite satisfactory for door frames. This joint is often used instead of a mortice and tenon joint.

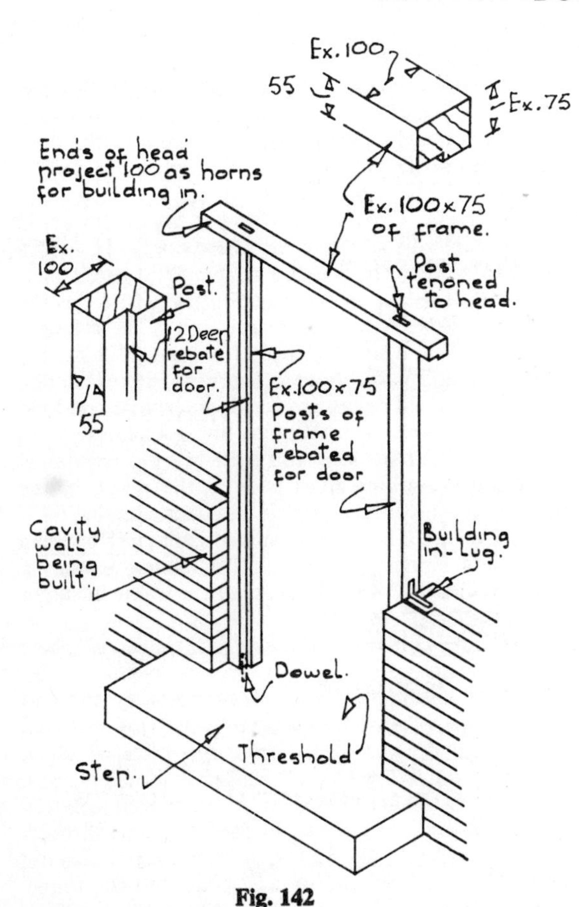


Fig. 142

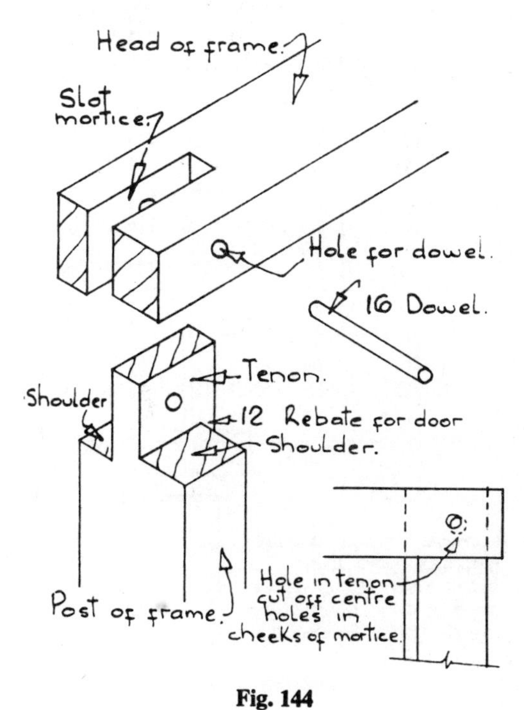


Fig. 144

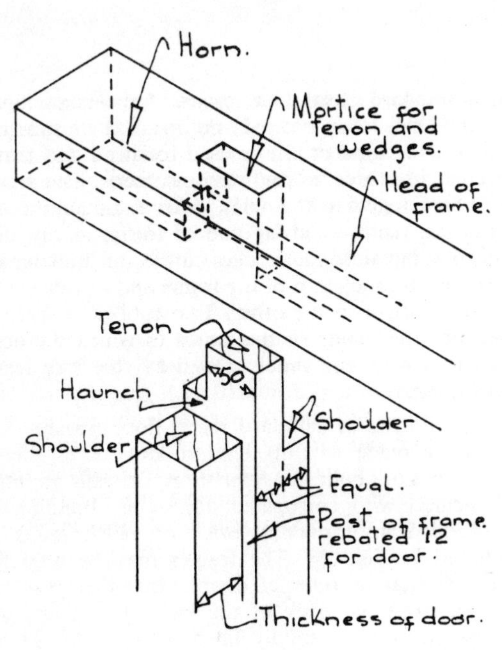


Fig. 143

**Dowels:** Door frames which do not have a timber threshold, and are built in over a concrete floor or concrete or stone threshold, are often secured to the floor by setting a dowel into the ends of the posts. The dowels are usually of mild steel 12 or 20 diameter and some 50 long. Half the length of the dowel is set into the end of the post and the other half either cast into or bedded in a hole in the concrete or stone. Door frames carrying heavy doors should be fitted with these dowels which fix the foot of the frame rigidly in position.

**British standard door frames:** The proposed metric door frame is made to suit standard metric doors as illustrated in Figs. 145 and 146.

Being mass produced in standard sizes these frames are cheaper than similar purpose made ones.

**Door sets:** As mentioned previously a door set comprises a door hung in its frame ready for delivery to site. The proposed standard metric door set frame is illustrated in Fig. 146.

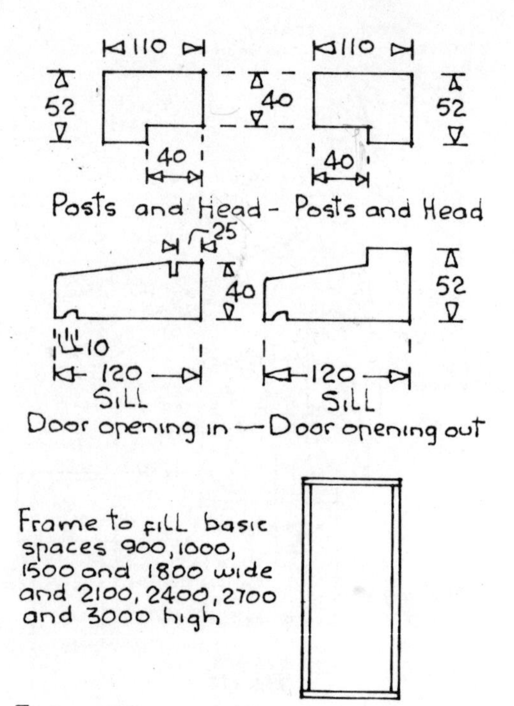


British Standard External door frame

**Fig. 145**

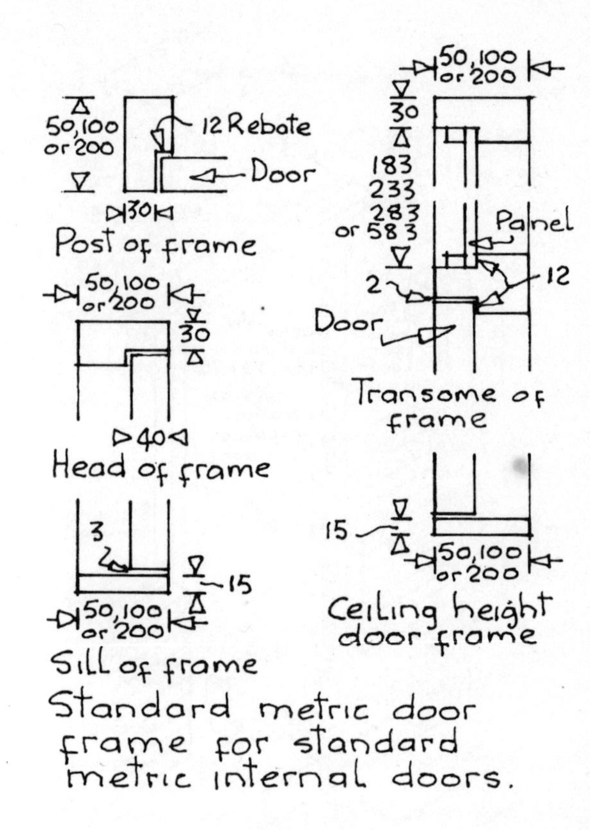


**Fig. 146**

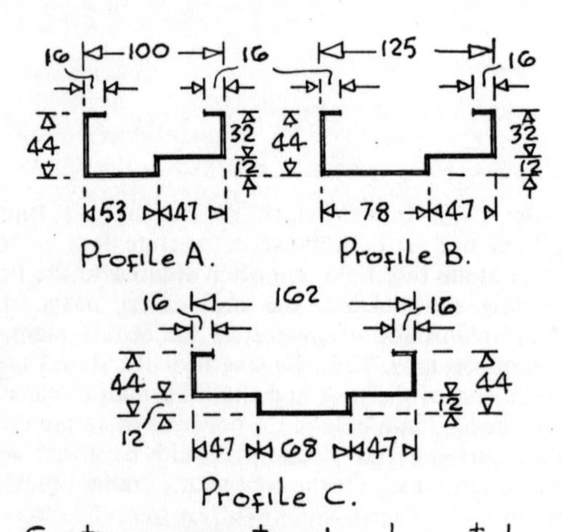


Sections of standard metal door frames.
Profile A. frames are made to suit external doors 6'-6" high and 2'-6", 2'-9", 3'-10" or 7'-0" wide and internal doors 6'-6" high and 2'-0", 2'-3", 2'-6" or 2'-9" wide.
Profile B and C frames are mad to suit internal doors 6'-6" high and 2'-0", 2'-3", 2'-6" or 2'-9" wide.

**Fig. 147**

**British standard metal door frames:** Are manufactured from mild steel strip pressed into one of three standard profiles. The same profile is used for head and jambs of frame. The three pressed steel members of the frame are welded together at angles. After manufacture the frames are rustproofed by one of the processes described for metal windows. Two loose pin butt hinges are welded to one jamb of the frame and an adjustable lock strike plate to the other. Two rubber buffers are fitted into the rebate of the jambs to which the door closes to absorb the shock caused by the door being closed heavily.

Fig. 147 illustrates standard metal door frames. The frames are made to suit standard door sizes. Metal door frames are built in as partitions or walls are built and secured with corrugated adjustable building in lugs bedded in horizontal brick or block joints as illustrated in Fig. 148. The frames may be used for either external or internal doors. In either case the space between the back of the metal frame and the wall or partition is solidly filled with mortar as the brick or block work is raised as illustrated in Fig. 149.

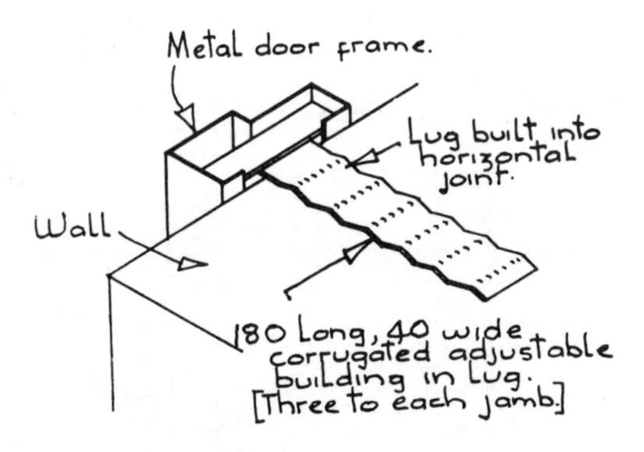


Fig. 148

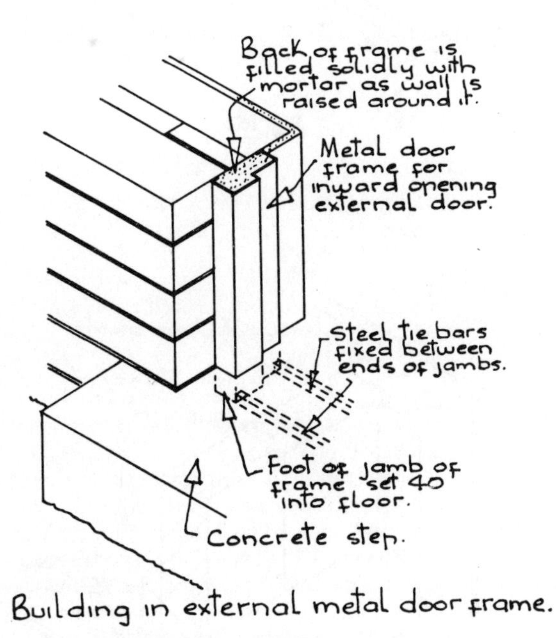


Fig. 149

When used for an internal door the metal frames are generally chosen so that they are wider than the finished thickness of the partition and this obviates the necessity for an architrave to mask the joint between plaster and frame. Some arrangements of metal door frames fixed internally are shown in Fig. 150.

The advantage of a metal door frame is that it will not shrink due to inadequate seasoning, as does a timber one. These frames are more expensive than similar standard wood frames and are not as yet greatly used in domestic buildings.

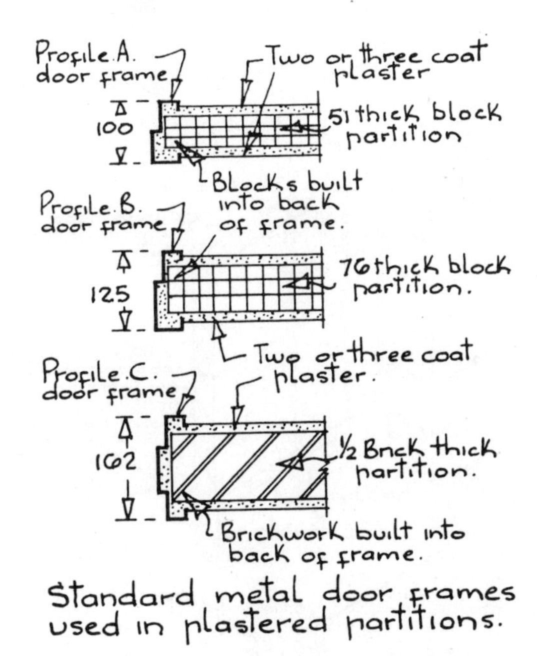


Fig. 150

**Thresholds (sills to doors):** The threshold or sill of a door is the surface below the door when it is closed. The threshold of an external door has to be designed to prevent wind and rain blowing in under the door. This is effected either by use of a rebated timber threshold or a metal water bar or both.

A timber threshold is usually of oak, 100 by 75, or sometimes 25 or 50 wider. It is rebated 12 deep for doors opening out, and weathered to throw water outwards. The posts of the frame are joined to it with mortice and tenon joints.

In Fig. 151 it will be seen that a throat is cut on the underside of the threshold to form a drip, and a water bar is set into grooves in the threshold and stone or concrete step below. The purpose of a throat and water bar was explained in Chapter 2. The floor inside is finished level with the top of the threshold so that there is no projection for people to trip over. The door shown in Fig. 151 is hung to open out of the building and the rebate in the threshold (sill) is generally sufficient to prevent wind and rain blowing in under the door. External doors are hung to open out so that they do not obstruct floor space inside the building when they are open. The disadvantage of hanging a door to open out is that if it is open during rain the top edge of it, which is not usually painted, becomes wet and the door may swell and be difficult or impossible to close. But if doors are made to open into buildings, a rebate in the threshold or sill will not prevent water getting in, as was explained with casements opening inwards. It is usual with inward opening doors to set a metal water

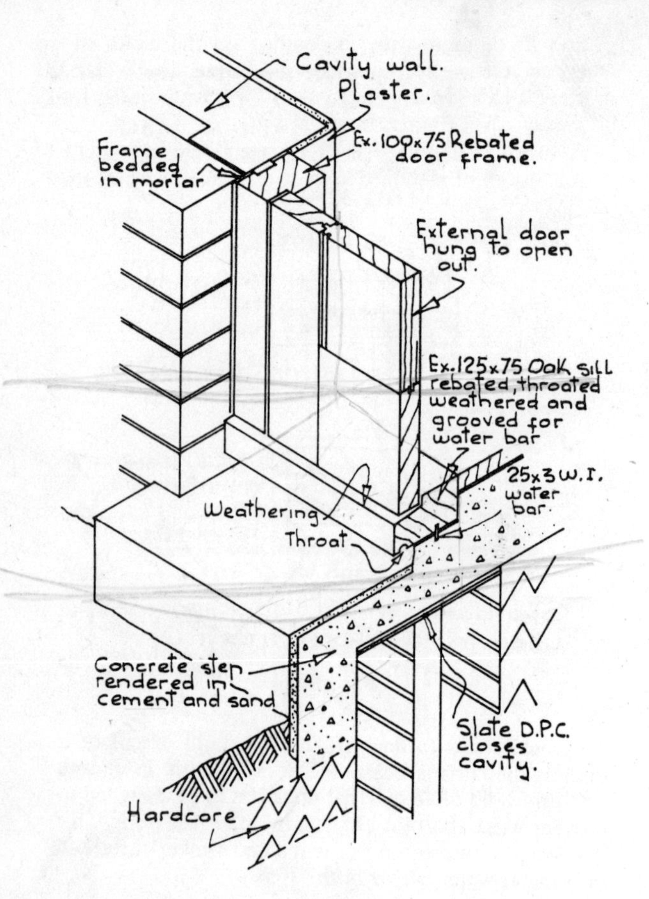


Fig. 151

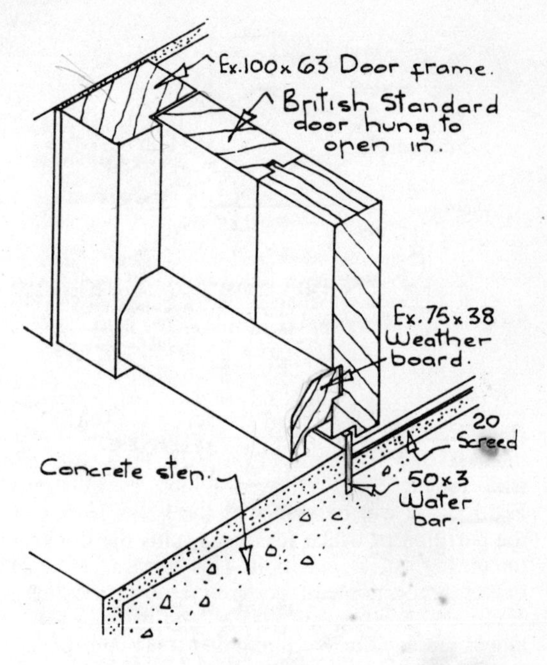


Fig. 152

bar into a wood threshold or sill or into a stone or concrete threshold, as illustrated in Figs. 152 and 153. The water bar is usually of galvanised steel or wrought iron, and either flat or "J" shaped. This latter section is used because it is thought that the tail of the "J" acts as a drip as illustrated in Fig. 154.

The disadvantage of these water bars, used with inward opening external doors, is that they are not very obvious and strangers trip over them.

**Weather board:** In addition to a water bar a timber weatherboard is often tongued into the bottom rail of inward opening doors as shown in Fig. 152. It throws water out away from the door.

Door frames are usually cut from softwood and thresholds from oak. The reason for this is that oak is more resistant to decay caused by water than is softwood and the threshold is usually more heavily saturated by rain than the other members of door frames.

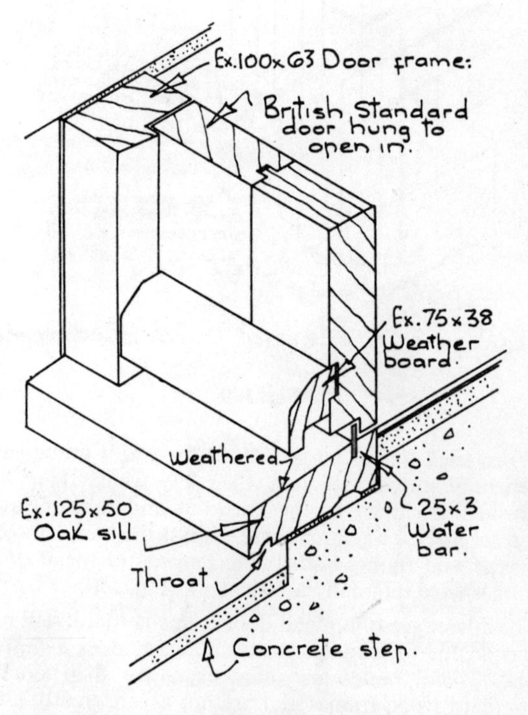


Fig. 153

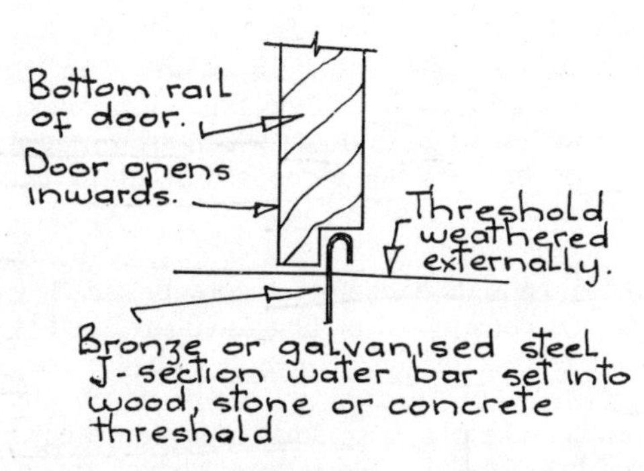


Fig. 154

**Door linings:** Door linings are principally used for internal doors. Boards 25 or 32 thick are jointed with tongued and grooved joints, as illustrated in Fig. 155. It will be seen that the lining is as wide as the partition in which it is fixed, plus the thickness of the plaster on the partition. This is arranged so that a timber architrave can be fixed around the lining to cover the junction of plaster and wood. The door hinges are fixed to the lining and a planted (nailed on) stop is fixed to the three members of the lining, to which the door closes. The planted stop is 12 thick

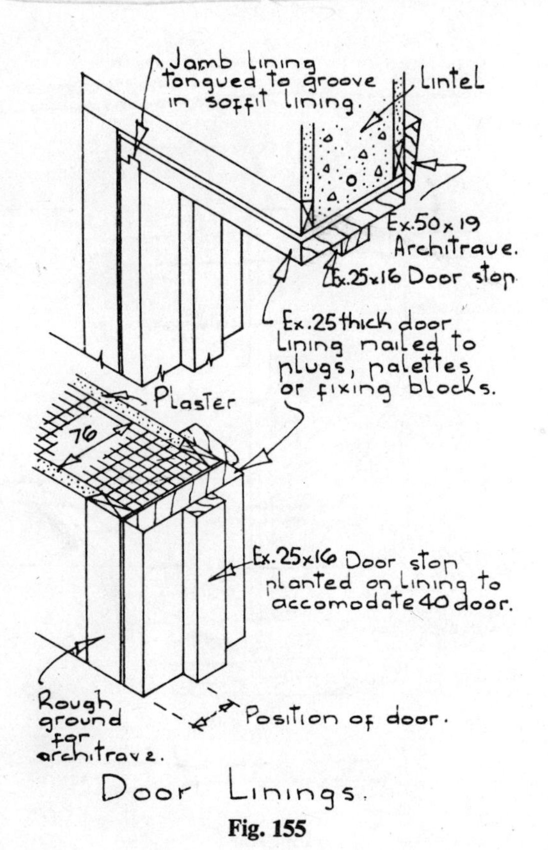


Fig. 155

and 25 or more wide, depending on the width of the lining. Generally the wider the lining, the wider the stop, as a narrow planted stop on a wide lining looks mean.

A range of standard door linings is manufactured to suit doors of standard sizes. Fig. 156 illustrates these.

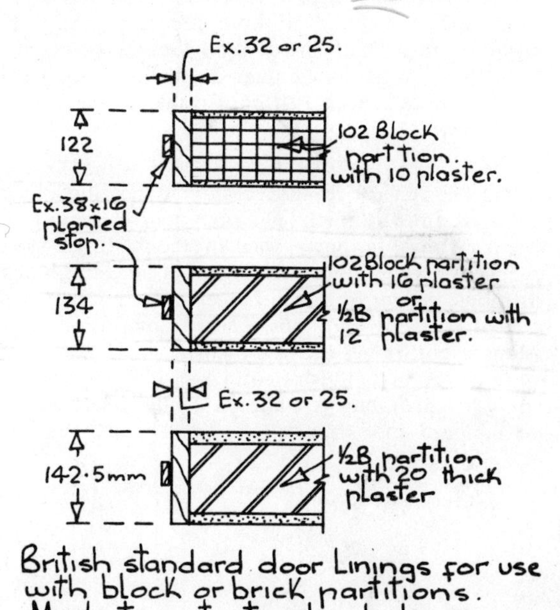


Fig. 156

**Fixing linings:** Linings are not built in. They are fixed in position after partitions are built, but before walls are plastered, so that the plaster can be finished up to them. It is not practical to nail linings directly to brick or dense block partitions because the large nails required for this would split and damage the lining. So plugs of wood, or rough timber grounds, or fixing blocks are used to which the lining can be nailed with small nails which do not split or damage them.

Plugs consist of wedge shaped pieces of wood which are driven into raked out joints between bricks or blocks. The lining is nailed to these wood plugs. This is not a particularly good fixing as the plugs have to be thin and nails driven into them split them, and they may come loose.

Grounds consist of lengths of small section timber which are nailed into the joints between bricks or blocks, the grounds being arranged either vertically or horizontally on reveals. If the grounds are damaged by nailing it does not matter as they will be hidden by the lining.

Rough timber grounds usually consist of 50 x 25 sawn batten, which are nailed to the reveals of the opening in the partition.

Purpose made fixing blocks are made from lightweight aggregate and cement in thicknesses to suit brick courses, that is 65.

They can be built into brick partitions at jambs every fourth or sixth course of bricks. They are sufficiently soft for small nails to be driven into them and sufficiently coarse textured to hold the nail firmly. If partitions are built of one of the lightweight concrete blocks (Vol. 1) the door linings can be nailed directly to them without much risk of damage to the lining or the blocks.

**Ceiling height door frames:** Non-load bearing block partitions up to 76 thick have such poor stability that door openings seriously weaken them. It is good practice to hang doors in openings in thin block partitions, in door frames put together so that they can be fixed at ceiling and floor. These frames are termed ceiling height frames because their posts are made the full floor height as illustrated in Fig. 157. A range of ceiling height frames is made of standard sections to suit standard door sizes.

## FLUSH DOORS

The present fashion in building is for plain surfaces devoid of decorative moulding which will collect dust. Hence the use of flush doors which are surfaced with sheets of plywood or hardboard fixed to either a skeleton frame or a solid core.

**Skeleton core flush doors:** A core of small section timbers is constructed as illustrated in Fig. 158. The main members of this structural core are the stiles, top, bottom and middle rail all of which are 30 x 80. Intermediate rails of smaller section are used as a rigid base for the plywood facings which are glued to the frame. The members of the skeleton core can be jointed with tongues cut on the ends of the rails and glued into grooves in the stiles as illustrated in Fig. 159.

It will be seen that a continuous groove is cut in the stiles into which the tongues on the stiles are glued and that the intermediate rails are similarly jointed to the stiles. The tongued joint illustrated in Fig. 159 is not so strong as the mortice and tenon or dowelled joints used for jointing rails to stiles in panelled doors.

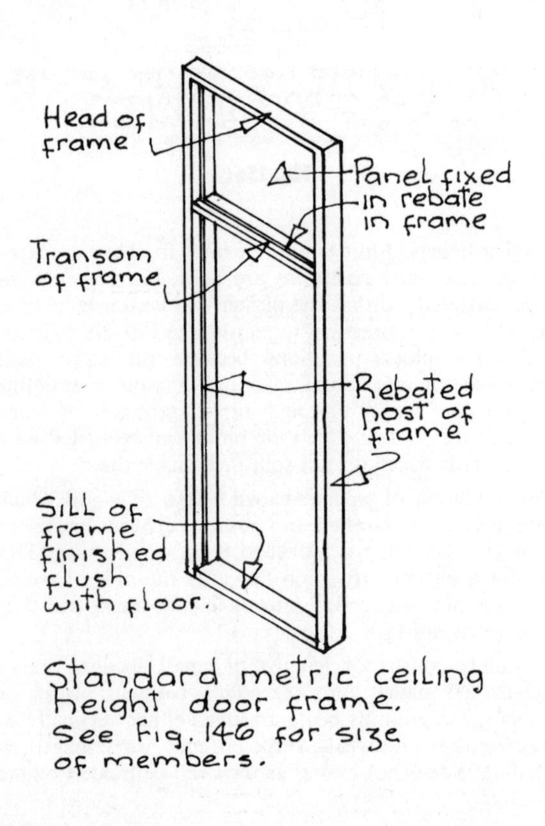


Fig. 157

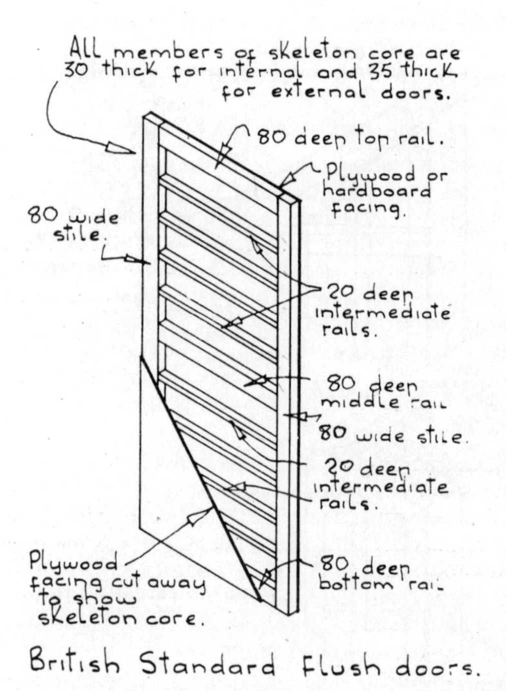


Fig. 158

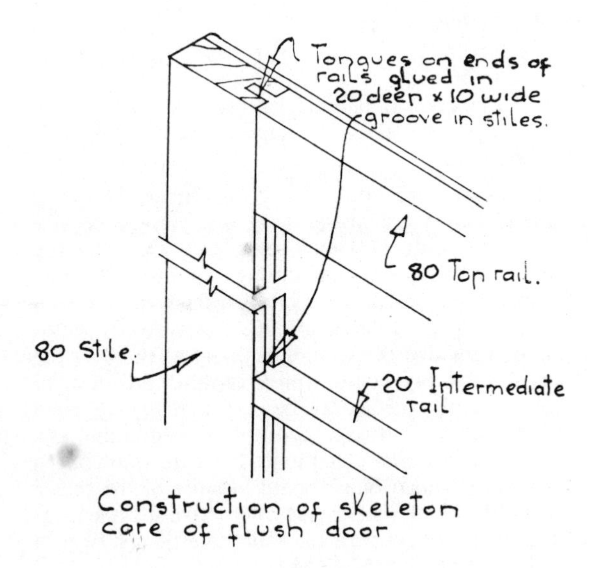


Fig. 159

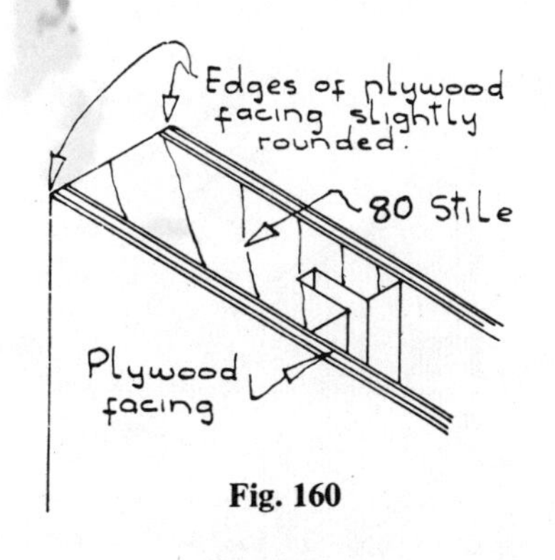


Fig. 160

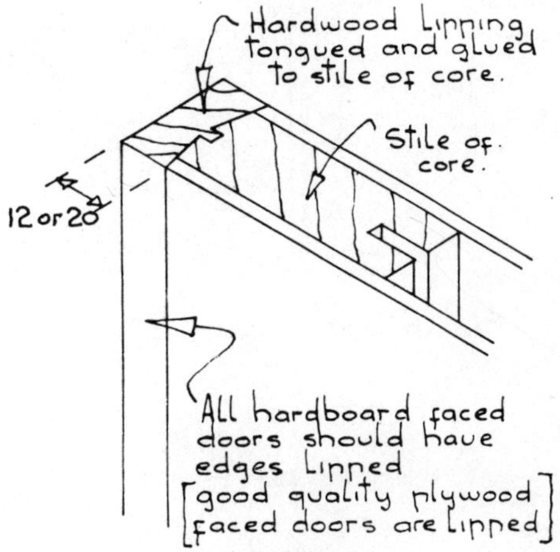


Fig. 161

But there is no necessity for very rigid joints between the members of the skeleton core of a flush door because the plywood or hardboard facings which are firmly glued to the core will effectively prevent any tendency for the door to sink out of square.

The plywood used as a facing for doors is usually three ply 6·5mm thick for external doors, and 5 for internal doors. Internal doors are sometimes faced with sheets of standard hardboard 5 thick. Plywood with face plies of some decorative hardwood is made so that the ply-boards can be polished or vanished to show the colour and grain of the wood. This class of plywood is used for facing the more expensive flush doors. One whole plyboard is used for each face of flush doors. The plywood is strongly glued to the skeleton core under pressure.

The cheaper flush doors have the edges of the plywood facings slightly rounded as illustrated in Fig. 160. But this does not provide a particularly neat finish to the edges of the door as it is not possible to cut the edges of plywood neatly due to the opposed grains of the plies. Many flush doors have a hardwood lip fixed to the two vertical edges of the door to mask the edges of the plywood as illustrated in Fig. 161.

The intermediate rails of skeleton core for flush doors are not always fixed horizontally. Fig. 162 illustrates another arrangement of rails in use today.

A range of standard flush doors is made in standard sizes and of standard thickness in the following sizes, 526, 626, 726 and 826 wide and 2·04 high.

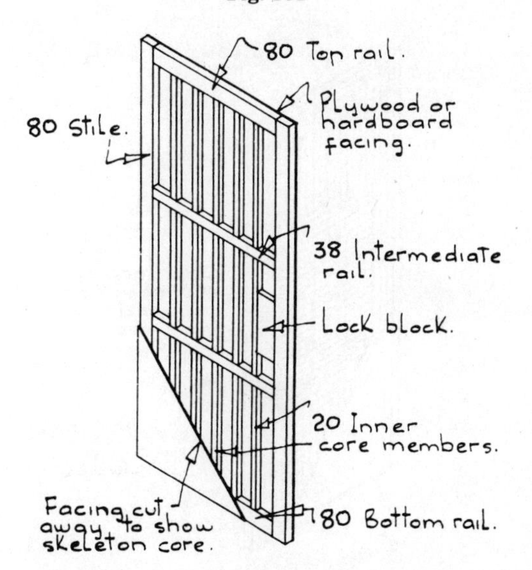


Skeleton core flush door [Not Standard] Rails tongued to grooves in stiles. Made in British Standard widths and height but 35 thick up to 726 wide.

Fig. 162

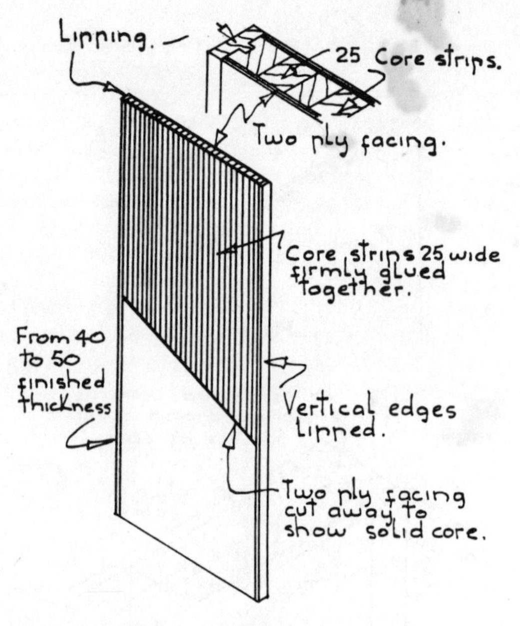


**Fig. 163**

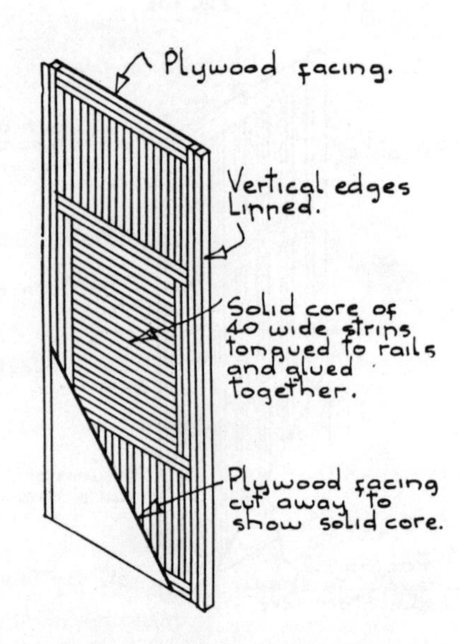


**Fig. 164**

**Solid core flush doors:** Plywood facings bonded to a skeleton core do not always remain absolutely flat and waves in their surfaces may be apparent particularly if the door is painted with a gloss paint. Also flush doors with skeleton cores do not appreciably exclude sound.

Flush doors with solid cores of timber, straw or synthetic board are made. They are more expensive than those with skeleton cores but their facings generally remain dead flat throughout their life. Fig. 163 illustrates the construction of a flush door with a solid core of 25 wide strips of timber solidly glued together with two ply facings both sides. In effect the door is a five ply board with a central core of solid strips. Two ply facings can be used here as the solid core acts as the centre (or core) ply in restraining any shrinkage in the plies glued to it. This type of plyboard is sometimes called block board because of the centre core of blocks of timber. An alternative arrangement of the strips of timber in the solid core of this type of door is illustrated in Fig. 164.

The makers of this particular door claim that the arrangement of the core strips produces a particularly rigid door unlikely to twist or lose shape. Providing the strips of wood used in a solid core are well seasoned, straight of grain and firmly glued together, their arrangement vertically, horizontally or both will not affect the strength and rigidity of the door. Flush doors are also made with a core of straw or wood shavings glued together, to which the three ply facings are glued. As these doors are neither a true solid core flush door nor a skeleton core flush door, they defy classification. It is claimed by the manufacturers that because of the great surface area of contact between the plywood facings and the core of straws or shavings, the surface of these doors remains truly flat. In practice their assertion has been justified. These doors are not as cheap as those with skeleton frames.

**Fire check flush doors:** Building regulations require certain doors to be able to resist the spread of fire for periods of half or one hour depending upon the size and type of building. For example, in a building divided into flats with a common staircase, the entrance doors to flats have to be constructed as fire check doors. These doors are constructed with a skeleton core to which four sheets of 9·5mm thick plasterboard are fixed in rebates in stiles and rails and over which the plywood facings are glued. Fig. 165 is an illustration of a fire check flush door to give half-hour protection. Plasterboard is fixed behind the plywood facings because it has good resistance to damage by fire. It consists of a core of gypsum plaster which has been cast between sheets of thick paper. The flush doors which are constructed to give one hour's protection have a skeleton core and plasterboard protective infilling, with sheets of asbestos wall board, or mill board, below the plywood facings. These doors are manufactured in the following standard sizes: 2·04 high and 826 or 726 wide. So that fire will not easily spread around the edges of these doors they are hung in frames with specially deep rebates. Fig. 166 illustrates the standard fire check door frame.

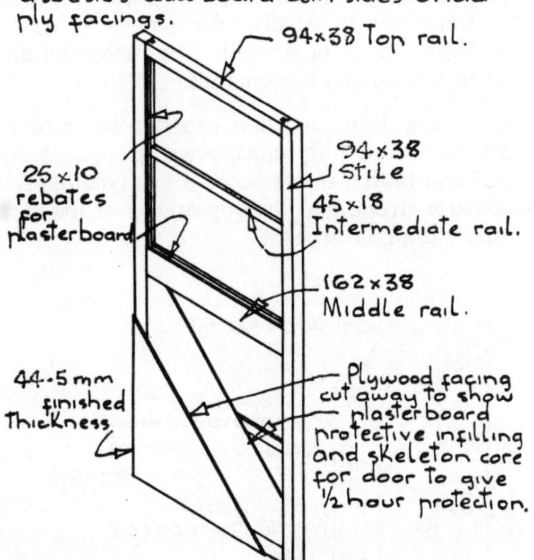


**Fig. 165**

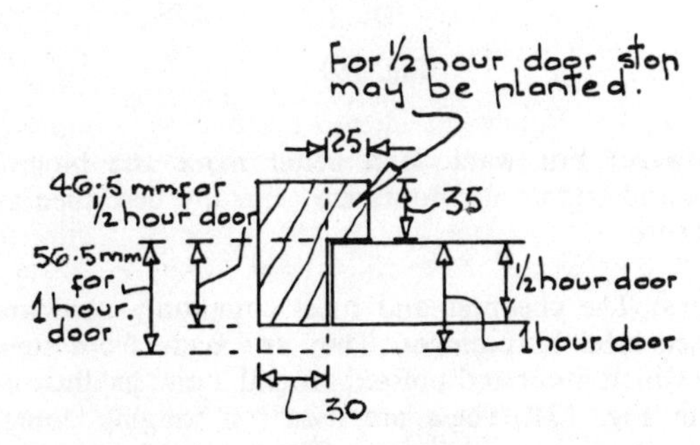


**Fig. 166**

**MATCHBOARDED DOORS (Battened doors)**

Up to the twentieth century most timber was cut by hand and the cheapest type of door that could then be made consisted of timber boards nailed to timber ledges as illustrated in Fig. 167. Today timber for most doors is cut, planed and jointed by machine and a standard size panelled or flush door is generally cheaper than one made of boards nailed to ledges. Consequently doors made with boards and ledges are rarely used. The disadvantage of a ledged and boarded door is that the nails securing the boards to the ledges are not very effective in preventing it from sinking out of square. Strap hinges were used to both hinge and strengthen these doors. They are used mostly for sheds and outhouses.

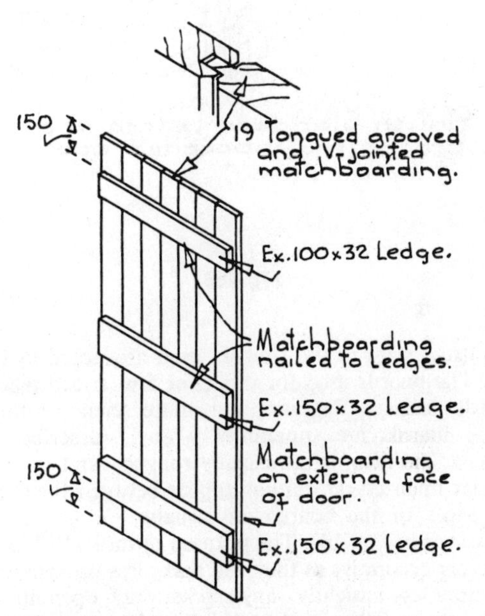


**Fig. 167**

**Ledged and braced matchboarded door:** A ledged and matchboarded door can be made reasonably rigid, so that it will not sink, by fixing timber braces diagonally between the ledges, as illustrated in Fig. 168. Obviously the braces must be fixed so that they incline upwards from the hinged edge of the door, otherwise they will not resist sinking.

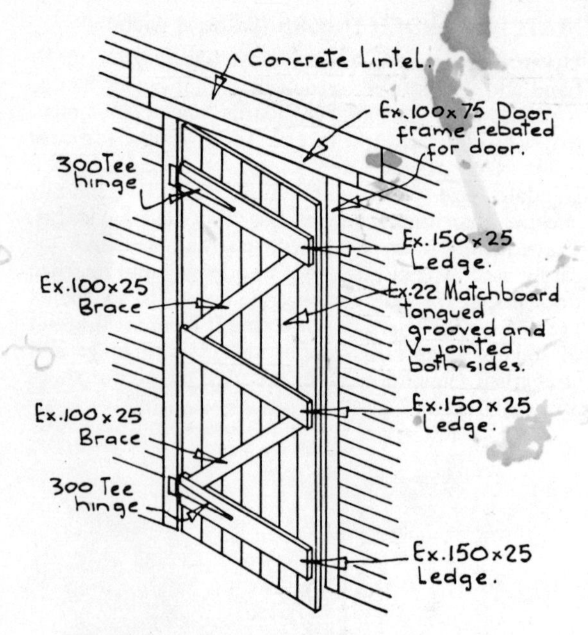


**Fig. 168**

The usual sizes of the timbers used are noted in Fig. 168. The boards used for this door are termed matchboards because they are not more than 19 thick. These boards are sometimes wrongly described as battens. The boards are usually tongued and grooved so that open cracks cannot appear between them and the edges of the boards are usually "V"-jointed as illustrated in Fig. 167. The purpose of these "V"-joints is purely decorative as they will make less obvious, and therefore less unsightly, any cracks which open up due to shrinkage. When used as an external door the matchboard face faces outwards. These doors are mainly used for sheds and garages.

**Framed and braced matchboarded door:** If a door is constructed of matchboarding fixed to a timber frame joined with mortice and tenon joints it will strongly resist any tendency to sink and if in addition braces are fixed between the rails it will be stronger still. This type of door is illustrated in Fig. 170.

These doors are mostly used as external doors with the matchboarding facing externally. The boards are fixed between the stiles and top rail and so that the boards can be run from the top rail to the bottom of the door, the middle and bottom rails are thinner than the stiles and rails.

The battens are tongued into the edges of the stiles as illustrated in Fig. 170. The top rail is tenoned into stiles as described previously for panelled doors Because the middle and bottom rails are less thick than the stiles "barefaced tenons" are cut on them, as illustrated in Fig. 169.

If a normal tenon were cut on the ends of these rails it would be too thin to make a strong joint. The tenons on the rails are glued and wedged in the mortices in the stiles. Large doors usually have pinned mortice and tenon joints for extra strength. The matchboarding is nailed to middle and bottom rails.

Large garage doors and doors to factories and warehouses are commonly constructed as matchboarded framed and braced doors because this type of door is particularly strong and the appearance of the door is not of great importance.

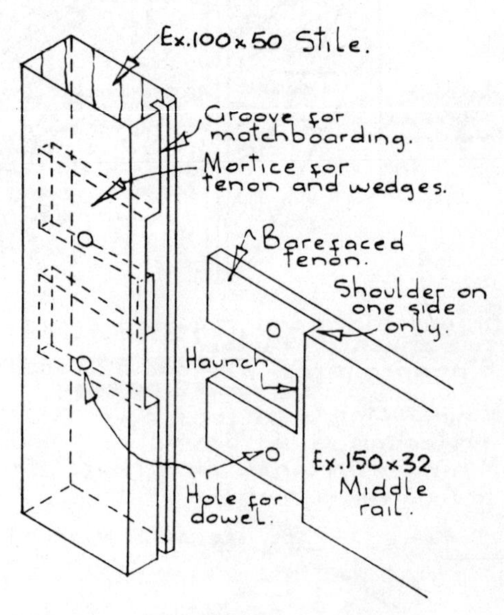


**Fig. 169**

**Hardware:** For want of a better name the hinges, locks and latches for doors are generally described as hardware.

**Hinges:** The cheapest and most commonly used are pressed steel butt hinges. They are made from steel strip which is cut and pressed around a pin, as illustrated in Fig. 171. These are used for hanging doors, casements and ventlights. The pin of standard steel butt hinges is fixed inside the knuckle. Loose pin pressed steel butt hinges are now manufactured. Their advantage is that by taking out the loose pin the door can be taken off its hinges whereas with standard steel butts a door can only be taken off by unscrewing its hinges from the frame.

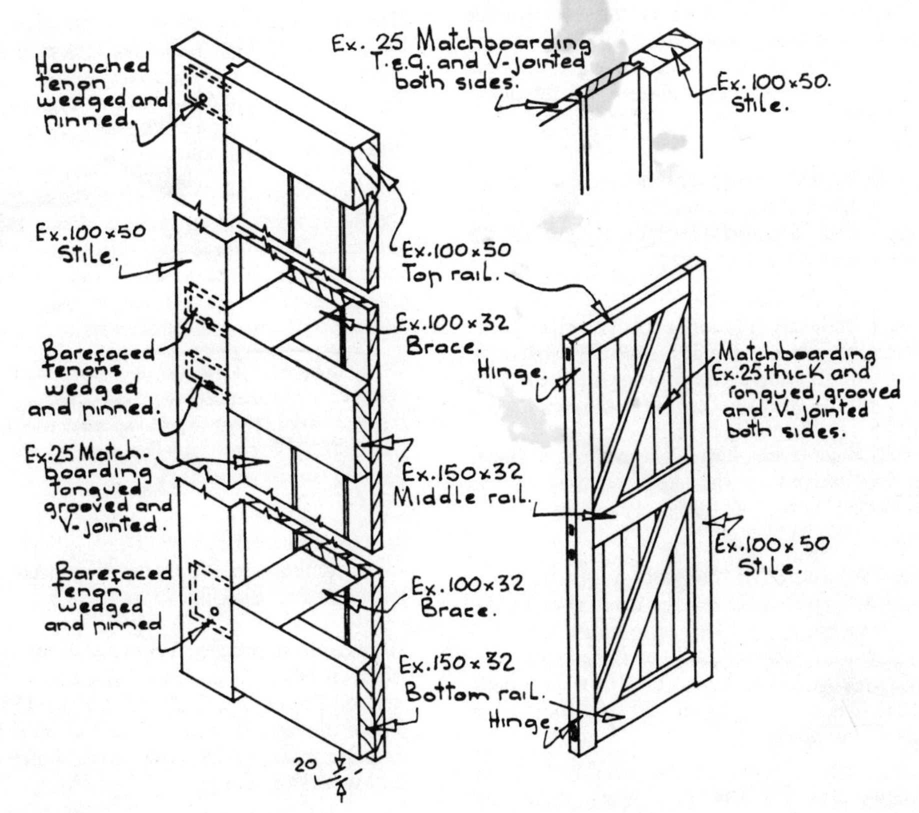
Haunched tenon wedged and pinned.
Ex. 25 Matchboarding T.&G. and V-jointed both sides.
Ex. 100×50. Stile.
Ex.100×50 Stile.
Ex.100×50 Top rail.
Ex.100×32 Brace.
Barefaced tenons wedged and pinned.
Hinge.
Matchboarding Ex.25 thick and tongued, grooved and V-jointed both sides.
Ex.25 Match-boarding tongued grooved and V-jointed.
Ex.150×32 Middle rail.
Ex.100×50 Stile.
Barefaced tenon wedged and pinned
Ex.100×32 Brace.
Ex.150×32 Bottom rail.
Hinge.
20
British Standard Framed and braced matchboarded door.
Standard sizes-

**Fig. 170**

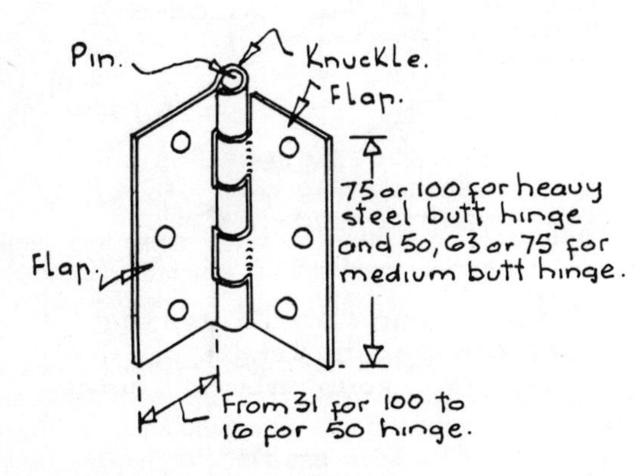
Pin.
Knuckle.
Flap.
75 or 100 for heavy steel butt hinge and 50, 63 or 75 for medium butt hinge.
Flap.
From 31 for 100 to 16 for 50 hinge.
Pressed steel butt hinges.
Are specified by length of flap as:
100 or 75 Heavy steel butt hinges and
75, 63 or 50 medium steel butt hinges.
Used for doors and casements.

**Fig. 171**

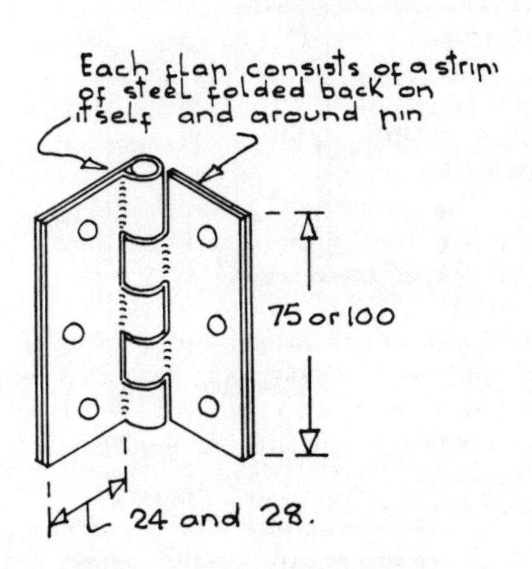
Each flap consists of a strip of steel folded back on itself and around pin
75 or 100
24 and 28.
Double pressed heavy steel butt hinge.
Used for heavy doors.

**Fig. 172**

**Double pressed steel butt hinge:** These hinges are made of two strips of steel each folded back on itself around the pin as illustrated in Fig. 172. They are stronger than ordinary steel butt hinges and are used for heavy doors.

**Cast iron butt hinges:** These are heavier and more expensive than steel butts of similar size and shape but have a longer useful life, as the bearing surfaces of the knuckles are more resistant to wear.

**Brass butt hinges:** These are more expensive than either steel or cast iron hinges. Used mainly for hanging decorative hardwood doors where the contrast of the colour of the brass and the wood makes for a pleasing effect.

**Steel skew butt hinge (rising butt):** The bearing surfaces of the knuckles are cut on the skew, so that, as the hinge opens, one butt rises. Fig. 173 illustrates a typical hinge.

These hinges are used for hanging doors and are fixed so that the flap screwed to the door rises as the hinge opens. The purpose of this is to cause doors to rise as they open to ride over, and so reduce wear on carpets. This type of hinge is also used for fire check doors, so that they tend to be self-closing due to the closing effect of the hinge.

**Steel tee hinges:** These consist of a rectangular steel flap and a long tail with knuckles around a pin as illustrated in Fig. 174. The flap is fixed to the frame

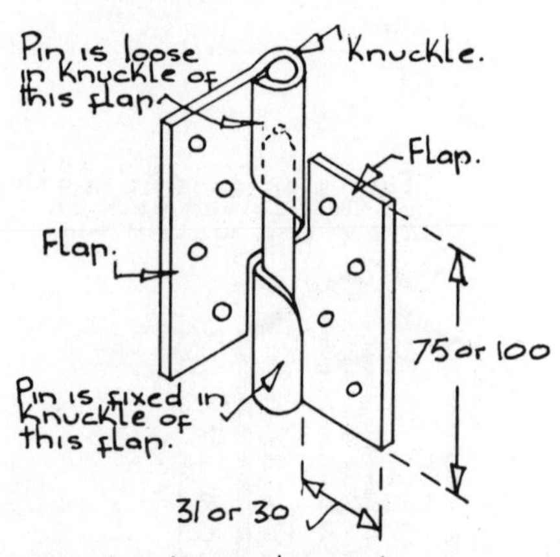


Fig. 173

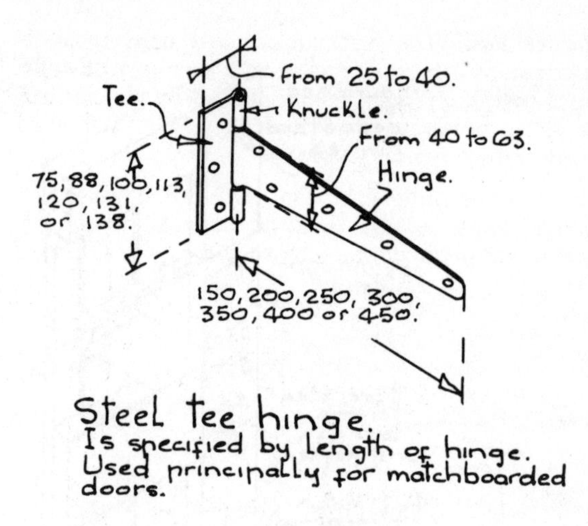


Fig. 174

and the tail to the door. These hinges are used mainly for matchboarded doors, as they assist in bracing the ledges against sinking.

**Hook and band hinges:** These consist of a rectangular steel plate in which a pin is fixed and a steel band folded around the pin as illustrated in Fig. 175.
These hinges are made of heavier steel than tee hinges and are used for hanging heavy doors such as garage and workshop doors.

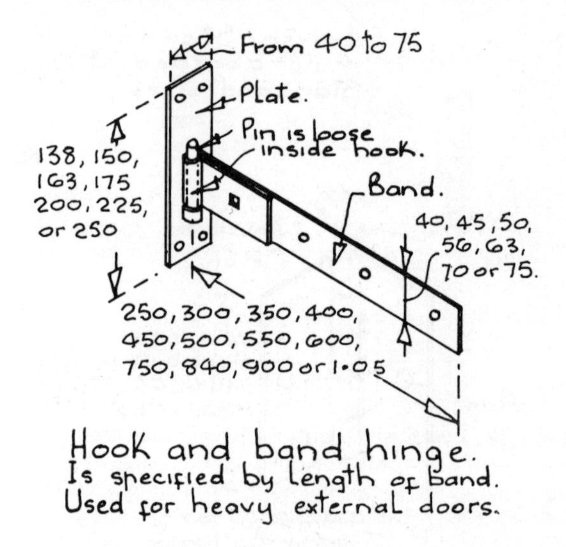


Fig. 175

**Latches and locks:** The word latch is used to describe any wood or metal device which is attached to a door to keep it closed and which can be opened by the movement of a handle, lever or bar. A lock is any device of wood or metal attached to a door which can be used to keep it closed by application of a loose key.

**Mortice lock:** The mechanism most used today is described as a mortice lock and it comprises a latch and a bolt, the former being operated by handles and the latter by means of a loose key. Fig. 176 is an illustration of typical locks.

The locks illustrated in Fig. 176 are described as mortice locks because they are designed to fit in a mortice cut in the door, so that the lock case is hidden

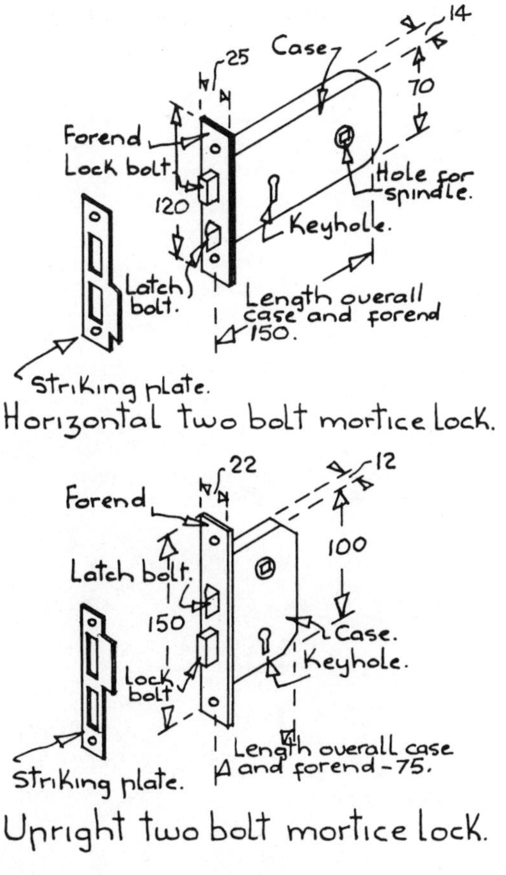

Horizontal two bolt mortice lock.

Upright two bolt mortice lock.

Fig. 176

**Rim lock:** Locks which are designed to be screwed to one face of a door are described as rim locks. They are not much in use today as they spoil the appearance of a door. Fig. 177 illustrates a rim lock.

**Cylinder night latch:** This consists of a metal cylinder which is housed in a metal case which is fixed to the inside face of a door. The cylinder fits in a hole in the door. The latch can be opened from outside by a loose key turned in the cylinder or opened by a knob from inside. The levers inside the cylinder are so arranged that only the key cut to fit a particular cylinder will open its latch. Fig. 178 is an illustration of a typical cylinder night latch. These latches are commonly used for front doors to houses and flats.

**Mortice dead lock:** This consists of a case inside which is a single bolt which can only be operated by a loose key. A typical mortice dead lock is illustrated in Fig. 179. These locks are fitted to a mortice in the door and the lock bolt shoots into a hole in a lock plate fixed to the door frame. These locks are used in addition to cylinder night latches or locks for entrance doors to houses and flats because they are more difficult to force or prise open than cylinder night latches.

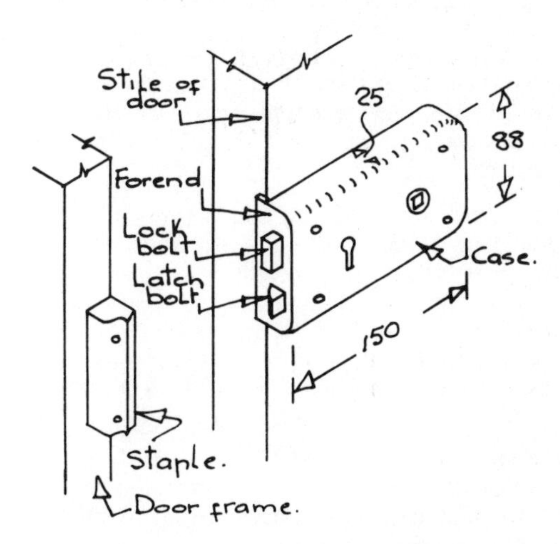


Fig. 177

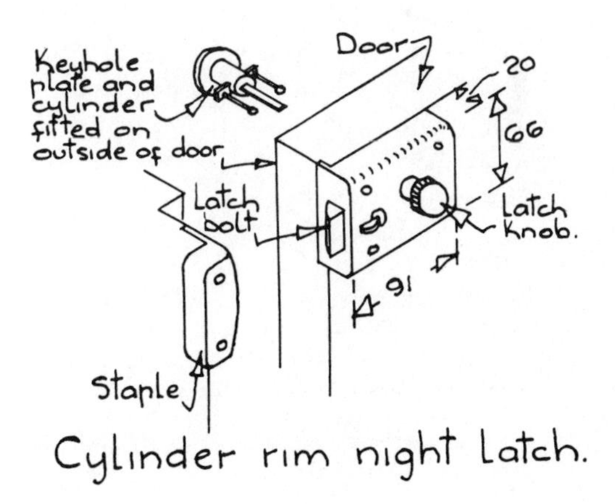


Fig. 178

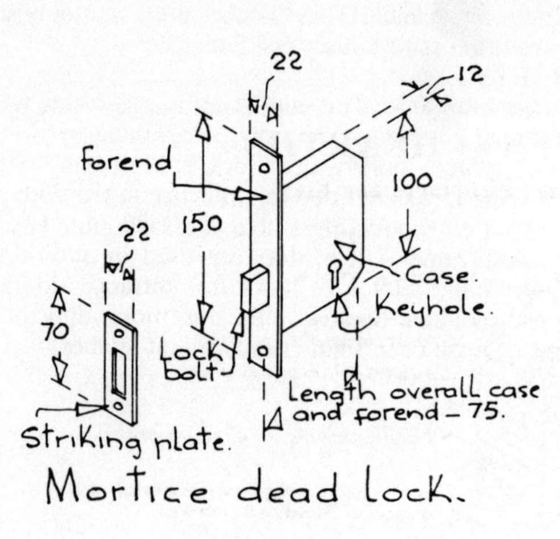


**Fig. 179**

**British Standards:**

No. 455. Locks and latches for doors.
No. 459. Part 1. Panelled and glazed wood doors.
Part 2. Flush wood doors.
Part 3. Fire-check flush doors.
Part 4. Matchboarded doors.

No. 1228. Door bolts.
No. 1245. Metal door frames.
No. 1331. Builders' hardware for housing.
No. 1567. Wood door frames and linings.

**British Standard Code of Practice:**

CP.151. Part 1. Doors and windows including frames and linings.

CHAPTER FIVE

# FIREPLACES—FLUES — HEARTHS

## SfB (56) Installations, heating: Fireplaces and flues

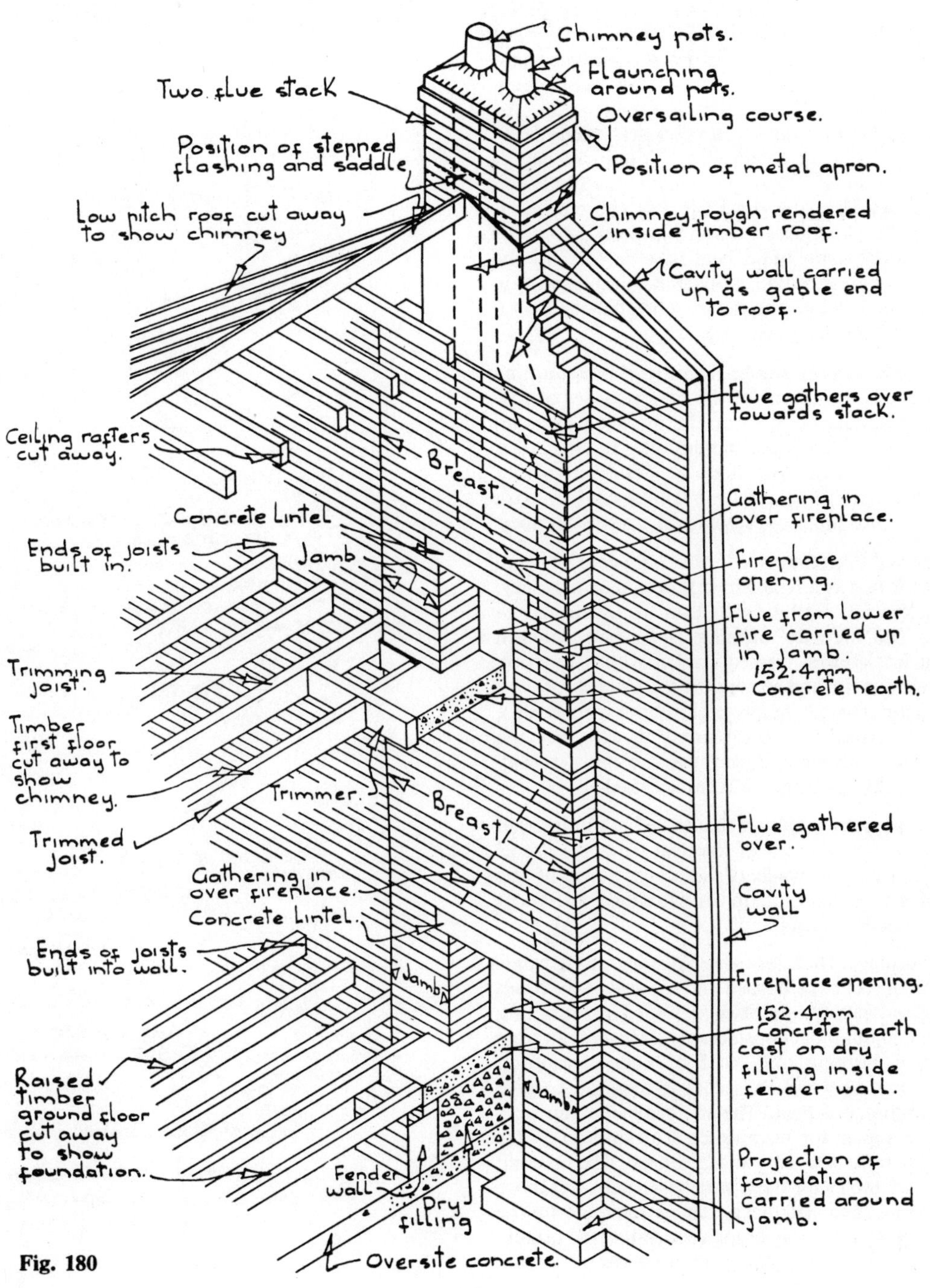
Chimney pots.
Flaunching around pots.
Two flue stack
Oversailing course.
Position of stepped flashing and saddle
Position of metal apron.
Low pitch roof cut away to show chimney
Chimney rough rendered inside timber roof.
Cavity wall carried up as gable end to roof.
Flue gathers over towards stack.
Ceiling rafters cut away.
Breast.
Gathering in over fireplace.
Concrete Lintel
Ends of joists built in.
Jamb
Fireplace opening.
Flue from lower fire carried up in jamb.
152·4mm Concrete hearth.
Trimming joist.
Timber first floor cut away to show chimney.
Trimmer.
Breast
Flue gathered over.
Trimmed joist.
Gathering in over fireplace.
Cavity wall
Concrete Lintel.
Ends of joists built into wall.
Jamb
Fireplace opening.
152·4mm Concrete hearth cast on dry filling inside fender wall.
Jamb
Raised timber ground floor cut away to show foundation.
Fender wall
Dry filling
Projection of foundation carried around jamb.
Oversite concrete.

Fig. 180

The open fire in which coal, coke or wood is burned is still favoured by the majority of people in this country in spite of considerable propaganda suggesting it is inefficient, laborious to stoke and clean, a major cause of fog and that it should be replaced by space heating (central heating) systems.

The conventional open fire is a very inefficient means of producing heat compared to a central heating boiler A great deal of the heat produced by burning coal, coke or wood in an open fire passes up the chimney and escapes from the building. A part of the fuel is often not burned at all and is thrown away with ashes. An average open fire is at best 30% efficient, which means that the fuel it uses to produce say 30 units of energy could have produced 100 units of energy in a 100% efficient appliance. But the open fire has the enormous advantage of looking cheerful and warm on a cold wet day even though some of its heat may be lost up the chimney, and a great deal of propaganda and legislation will be required before the last Englishman is driven from his armchair beside an open fire.

In cities and towns the smoke from open fires burning coal is unquestionably one of the major causes of fog. But there are smokeless fuels on the market which can be burned in open fireplaces and do not give off smoke. Many of the modern open fires are designed so that much of the heat given off by burning coal or coke finds its way into the room instead of up the chimney

The traditional open fireplace for burning coal or wood is usually formed as a recess in brickwork or masonry, projecting from a wall or partition as illustrated in Fig. 180. The fireplace opening is formed between projecting brick piers, termed jambs, whose function is to contain the fire and to support the brickwork over the fireplace opening. The projecting brickwork over a fireplace is termed the breast and it contains the flues through which smoke from the fire finds its way out of the building through a chimney stack.

Where ever possible fireplaces should be built in internal partitions so that as much of the heat from the fire as possible, warms the inside of building. Some of the heat from a fireplace built in an external wall is lost to the air outside the building.

**Fireplace openings:** The size of a fireplace opening depends on the type of fuel to be burned and the size of the room that the fire heats. A fireplace designed for burning only wood should be considerably larger than one designed for burning only coal or coke because a greater volume of wood than coal is required to give off a set amount of heat. The majority of open fires today are designed for burning coal or coke in grates which are at most 600 wide and 225 deep. The opening between brick jambs for these grates is usually not more than 750 wide and 350 deep. The brickwork in the chimney breast over the fireplace opening is usually supported by a pre-cast concrete lintel 140 deep, ½B wide and built into brickwork at jambs not less than 75 for openings up to 750 wide. Fig. 181 is a view of a fireplace opening showing the lintel. It will be seen that the lintel

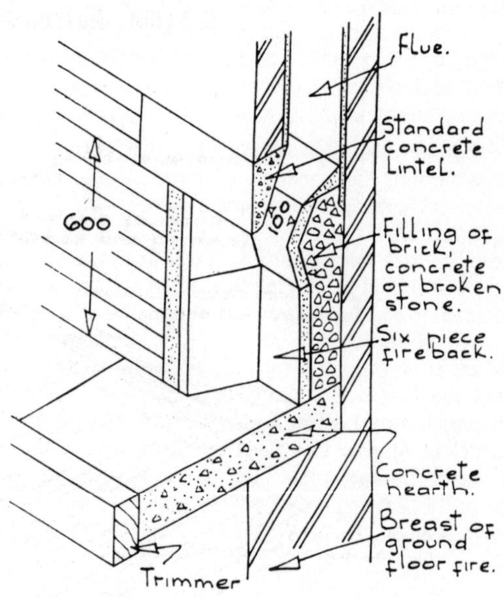


**Fig. 181**

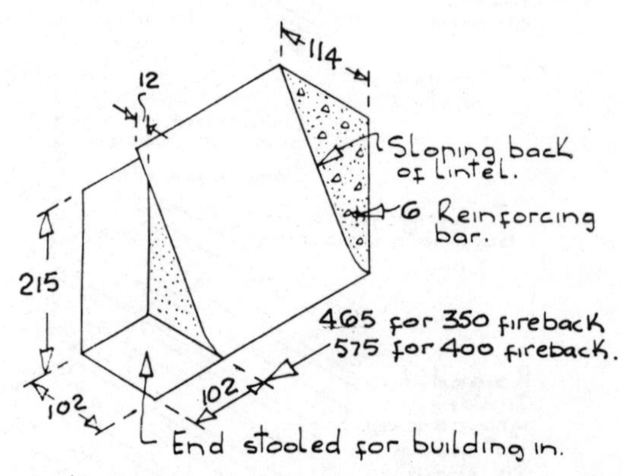


**Fig. 182**

is not cast with a normal rectangular section but has its inside face sloping, as shown in Fig. 182. The purpose of this slope is to assist in directing smoke from the fire up to the flue. The fireback shown in Fig. 181 is built in to contain the heat of the fire and the projecting knee at the back of it serves to form a narrow opening between the fire and the flue, termed the throat. The purpose of this throat is to restrict the amount of air drawn up the flue and so minimise the amount of heat that escapes up the flue.

The concrete lintel illustrated in Fig. 182 is the British Standard lintel. This section of lintel should be built in so that its sloping back forms a throat with the back of the firebrick as shown in Fig. 181 and to achieve this the firebrick back has to be built in as the chimney is built. But the firebrick may be stained and discoloured by mortar or damaged as the chimney above it is built and many builders prefer to fix the firebrick back after roofs are covered in. A rectangular lintel is then built in over the fireplace opening and later the fireback is set in position and a few courses of bricks built below the lintel as shown in Fig. 183. Details of sizes and materials of firebacks will be given later with details of grates and fire surrounds.

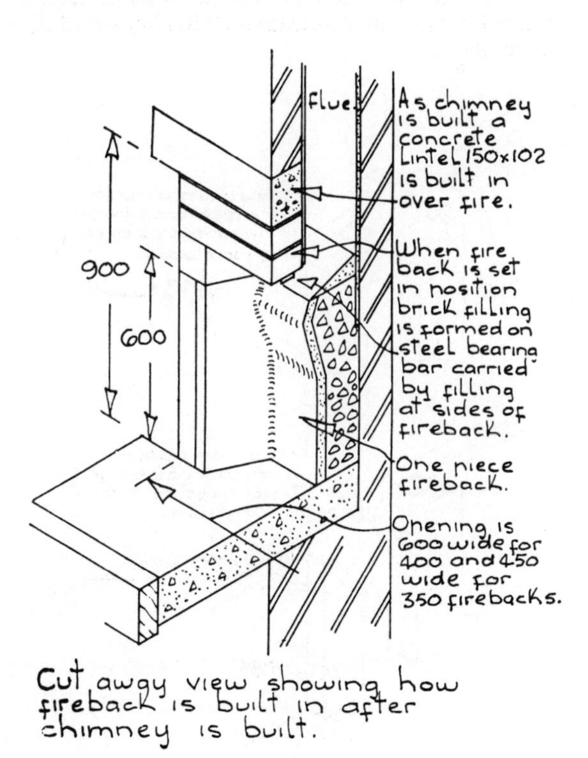


**Fig. 183**

**Jambs:** As has been stated the purpose of the brick jambs either side of a fireplace is to contain the fire and support the projecting brickwork in the chimney breast over the fireplace. These brick jambs should not be less than 1B wide on face, which is more than the minimum width required by The Building Regulations and G.L.C. By-laws to prevent heat from the fire setting timber at side of jambs alight.

One or both of the jambs of upper floor fireplaces has to be at least 2B wide on face to contain the 1B square flues carried up from fireplaces on lower floors. This is explained later under the heading Flues. The projection of the jambs from the face of the wall or partition in which they are built, depends on the difference between the thickness of the wall or partition and the thickness of the chimney breast which they support.

**Flues:** The Building Regulations and G.L.C. By-laws require every open fire to have a separate flue carried outside the building. The conventional flue for open fires is usually about 225 square and is formed in the brickwork of chimney breasts, jambs and chimney stacks. The Building Regulations require that flues be at least 177·8mm and G.L.C. By-laws at least 152·4 across at all points.

A 1B (225) flue is adequate for open fires burning coal or coke and is used because it is a convenient size of shaft to form in brickwork. Fireplace openings are wider than 225 and the brickwork at the sides of the opening has to be gathered in to the flue. Where there are two or more fireplaces over each other at different floor levels in a chimney the flues from the lower fireplaces have to gather over so that they run up in the jambs of upper fireplaces as illustrated in Fig. 184. The angle at which the gatherings in over a fireplace, and the angle at which flues gather over to run up in jambs of upper fireplaces, is usually about 53 degrees. This angle is determined by the bond in brickwork, namely ¼B, the slope of the flue being determined by ¼B off-sets in every brick course as illustrated in Fig. 185.

Brick flues are lined with pargetting which is a mixture of cement and sand in the proportions of one of cement to three of sand, or cement, lime and sand in proportion of 1:2:9. Pargetting is spread on the surfaces of flues, as they are built, to a thickness of about 12. The purpose of pargetting is to form a smooth faced lining which will not readily collect soot, will not obstruct the draught of the fire and can be cleaned when the flues are swept.

**Chimney breast:** The breast is the mass of brickwork over the fireplace and jambs and its function is to contain the flue or flues from fireplaces below.

The Building Regulations require the flues to be surrounded by at least 101·6mm thickness of bricks or blocks. The projection of the chimney breast is therefore determined by the difference between the thickness of the wall or partition in which it is built and the over-

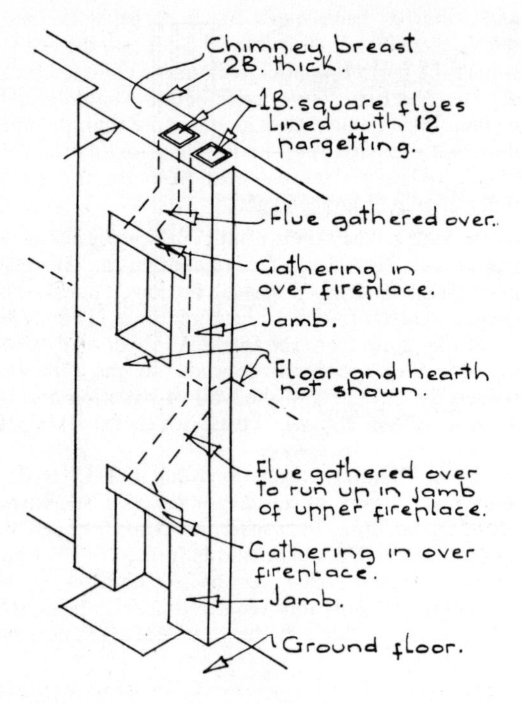


Fig. 184

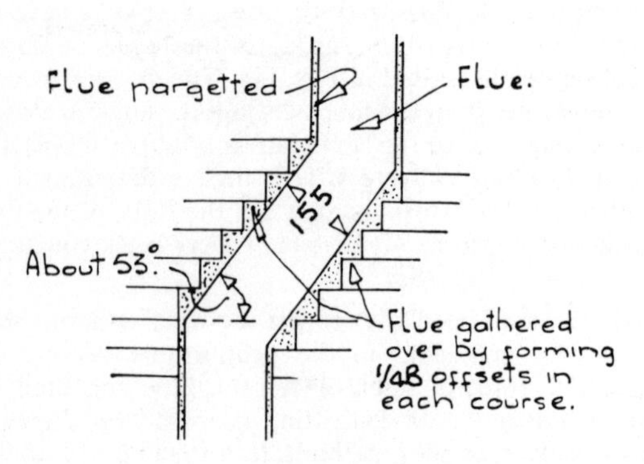


Fig. 185

all thickness of the breast, which is generally at least 2B. (1B flue plus ½B enclosing brickwork). This is illustrated in Fig. 186. Obviously the thicker the wall the less the projection of the breast and in a wall 2B thick the breast need not project from the wall at all. The overall width of the chimney breast depends on the width of the fireplace below it and the number of flues built into the breast. This is illustrated in Fig. 187.

It will be seen from Fig. 187 that when there are several fireplaces one over the other on different floors, the breast over the top fireplace has to be quite wide so that the flues from lower fireplaces can run up in the jambs at the side of the fire. The breast can be built the same width on each floor as illustrated in Fig. 187(b) or it can be made wider at each floor to allow for one extra flue in the jambs as in Fig. 187(c). There is some slight saving in cost in doing this as opposed to the arrangement as shown in Fig. 187(b). Which of the two types of construction is used is mostly a matter of taste. Many prefer the wider jambs at lower floor levels as they make for a more imposing fireplace.

The chimney breast and jambs do not have to be built projecting into the room which the fire heats. They can be formed as a projection outside the external walls or on the opposite side of partitions as illustrated in Figs. 188 and 189.

The advantage of this in external walls is that the breast and jambs do not take up useful floor space. The disadvantage of this arrangement is that the chimney is exposed on three faces to cold outside air which will cause excessive loss of heat from the fire and may cause the flues to cool to such an extent that the draught from the fire is reduced. The thickness of brickwork between flues in external walls and external face of breast is often increased to 1B to reduce cooling of the flue.

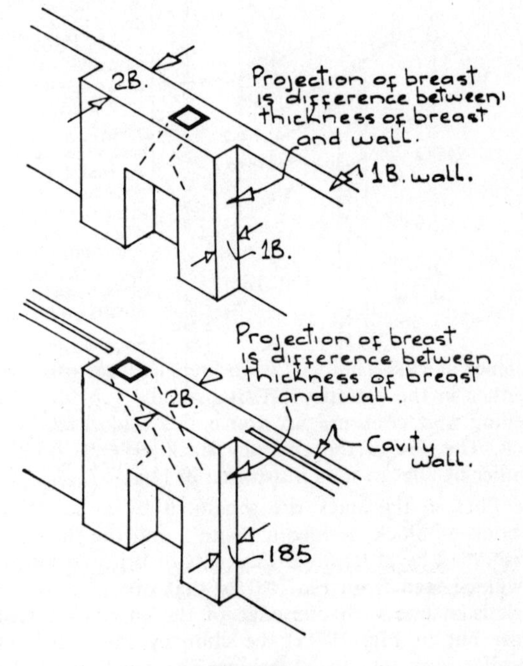

2B.
Projection of breast is difference between thickness of breast and wall.
1B. wall.
1B.
Projection of breast is difference between thickness of breast and wall.
2B.
Cavity wall.
185

Fig. 186

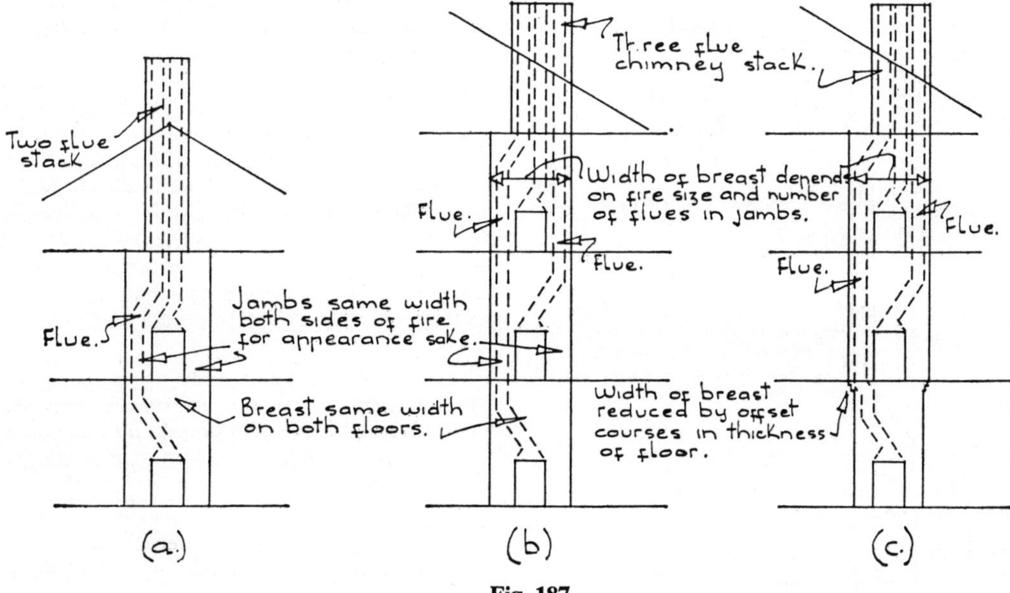


Fig. 187

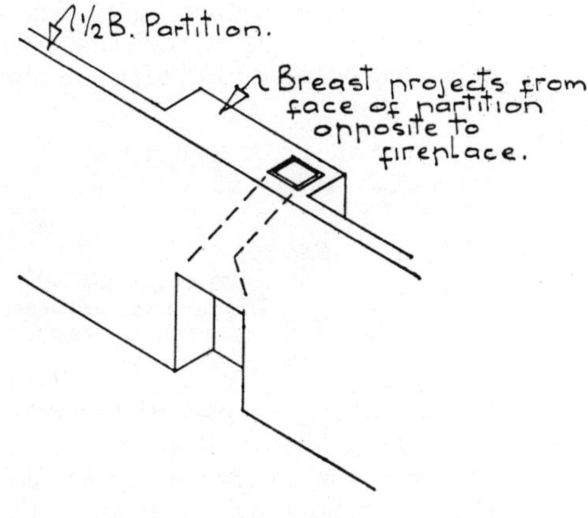


Fig. 188

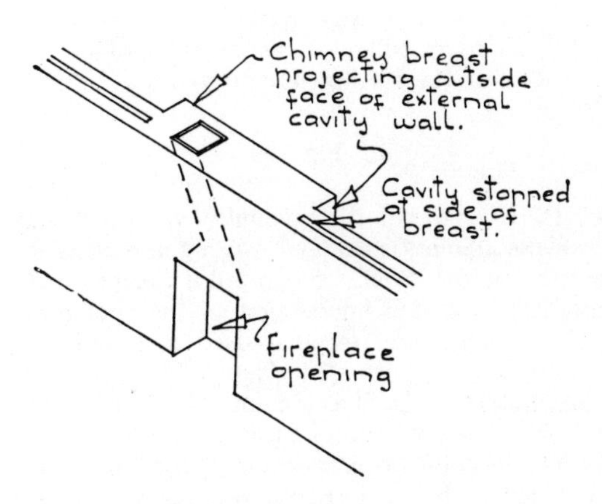


Fig. 189

**Chimney stacks:** The flues from fireplaces are gathered together in the chimney breast over the top fireplace opening and continue up above the roof level in a stack. The size of the chimney stack depends on the number of flues in it as illustrated in Fig. 187.

The flues in the stack are separated by at least $\frac{1}{2}$B of brick or block, termed a "with", and the flues are surrounded by at least $\frac{1}{2}$B thickness of brick or block. It will be seen from Fig. 187(b) that one face of the stack is in line with one edge of the chimney breast below but in Fig. 187(a) the chimney stack is built centrally over the breast below. The position of the stack in relation to the breast below is a matter of taste depending on where it will look best in relation to the ridge of the roof.

The least height of the stack above roof level is limited by building regulation. Fig. 190 illustrates the regulations diagrammatically. The flues in a chimney stack must be pargetted just as they are below. The maximum height of a stack is limited to not more than six times its least overall thickness.

Because a chimney stack is exposed to rain and frost, it should be built of sound, hard, well burned bricks or dense blocks or dense stone in cement mortar (1 part cement to 3 sand).

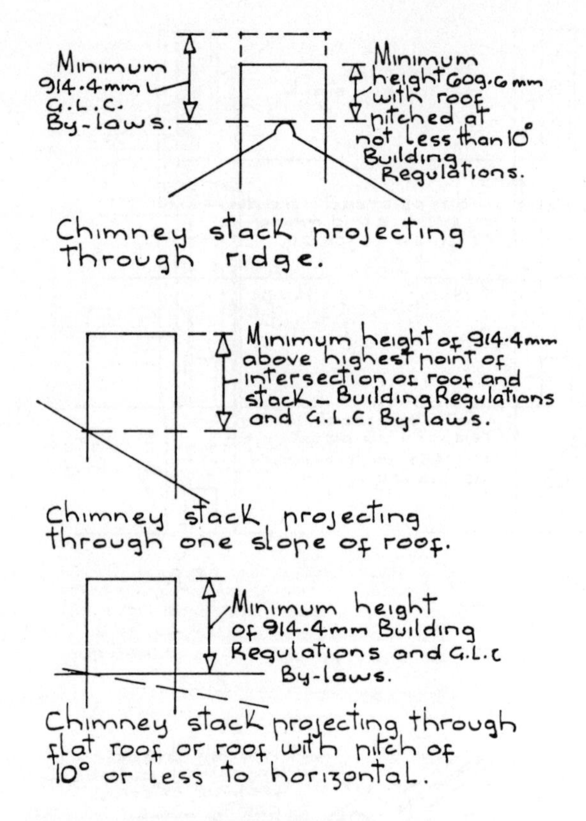


**Fig. 190**

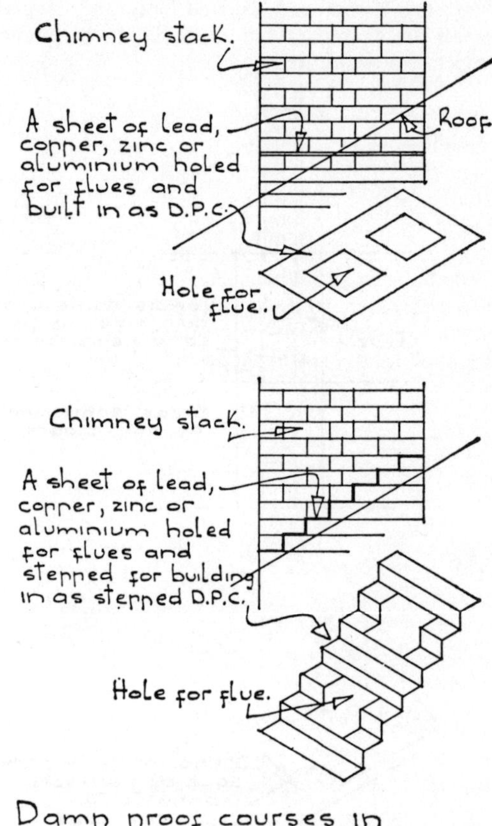


**Fig. 191**

**D.P.C.:** To prevent the possibility of water soaking down the chimney stack and causing dampness in the walls below roof level, a damp proof course should be built into it at the intersection of the roof and the stack. Usually sheet metal is used as a D.P.C. and commonly one sheet of 1·14mm copper is used in which holes for the flues are cut. The D.P.C. may be built in to the horizontal joint in the stack in line with the middle point of intersection of roof and stack as illustrated in Fig. 191. But it will be seen from Fig. 191 that there is a small triangular area of brickwork between the D.P.C. and the roof, and dampness may soak down from this brickwork. The likelihood of this occurring is very slight. Alternatively a D.P.C. stepped to brick, block or stone courses may be used as shown in Fig. 191. The D.P.C. in chimney stacks rising through flat roofs is built into the horizontal brickwork joint at the level of the top of the skirting or flashing of the roof covering as illustrated in Fig. 192.

**Chimney pots—terminals:** The top courses of a chimney stack may be finished with oversailing courses as a decoration, or finished with brick on edge. Whatever arrangement is used is a matter of taste. Fig. 192 illustrates the method of setting the chimney pot in the stack.

There are a variety of chimney pots on the market, the principal types being illustrated in Fig. 193. The purpose of chimney pots is to provide a smooth outlet to the top of flues to allow the smoke to rise freely, and to raise the level of the outlet for smoke above the top of the stack whose height is restricted by regulation. The reason for this is to prevent what is known as down-draught. The expression down-draught describes the effect of wind blowing down a flue and forcing smoke back into the room below. Down-draught is usually caused by surrounding buildings or trees which are higher than the top of the stack and which, in deflecting wind cause it to blow down into the lower chimney. This is illustrated diagrammatically in Fig. 194. The intensity of downdraught depends on the relative position of the flue affected by it and nearby trees or buildings and the direction and intensity of wind. It is not possible to forecast the intensity of down-draught likely in the flues of new buildings because of the variability of the causes. For this reason a variety of "tall boy" pots and louvred pots and revolving terminals are manufactured to eliminate or minimise the effect of down-draught.

**Flaunching:** Chimney pots and terminals are bedded on the chimney stack and there is an area of horizontal surface around them. So that this surface does not become saturated by rain and damaged by frost it is protected by weathered flaunching around the pot as illustrated in Fig. 195. Flaunching is a mixture of washed coarse sand and cement (in the proportions of 3 to 1) and water, which is spread around the pots or terminals to keep them in position and finished off smooth and weathered to throw off water.

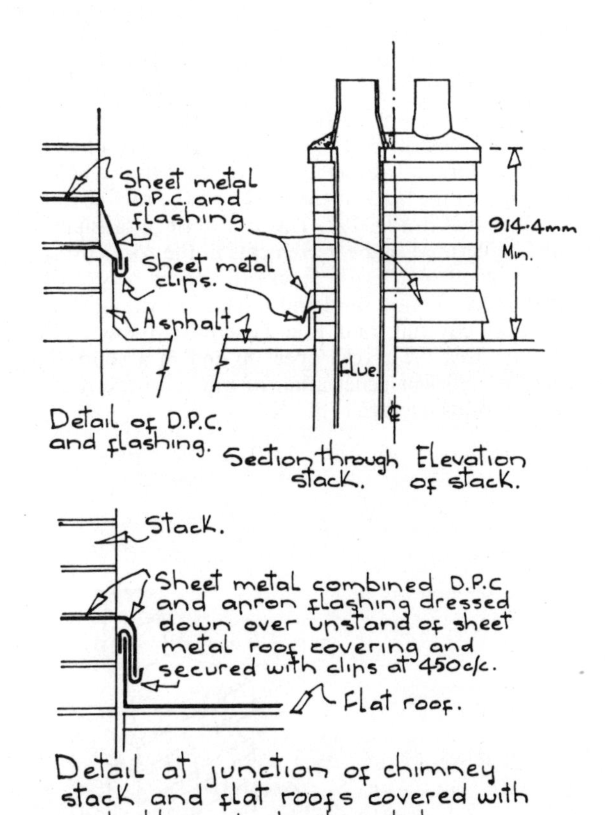


Fig. 192

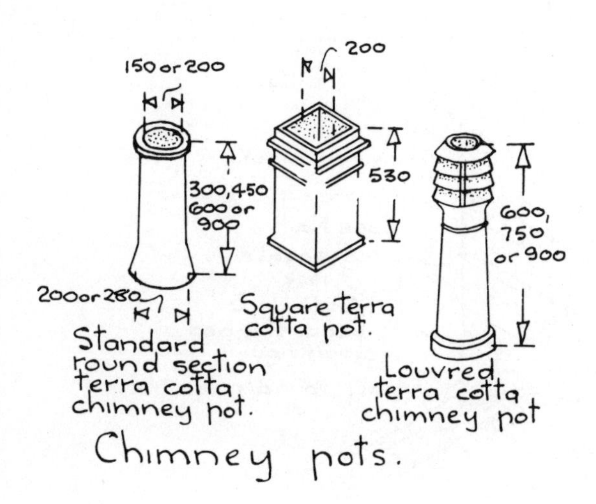


Fig. 193

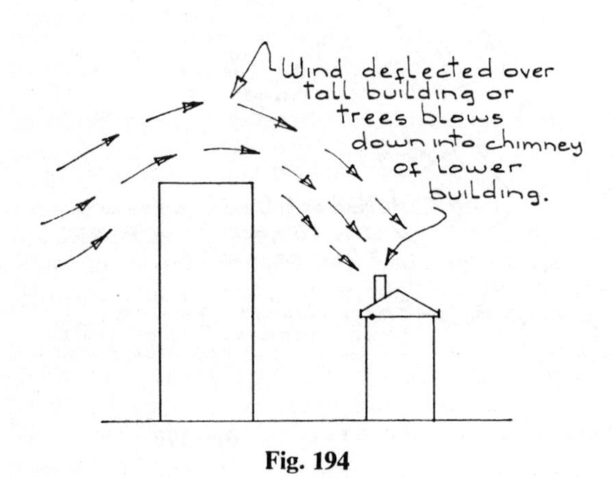


Fig. 194

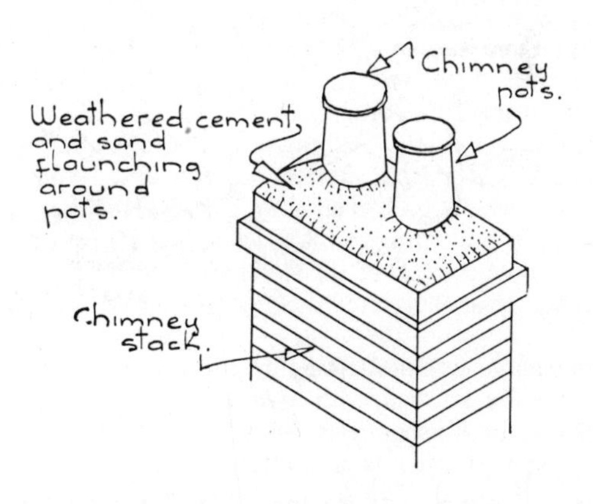


Fig. 195

**Weathering around stacks:** Where a brick stack rises through the surface of a pitched roof there is a gap between the faces of the stack and the roof covering. This gap must obviously be permanently covered or sealed to prevent water getting in. One way of doing this is to form a weathered fillet of sand and cement in the angle of the faces of the stack and the roof covering as illustrated in Fig. 197. This is a most unsatisfactory method because slight movements in the roof covering due to wind soon cause cracks to appear between the fillet of cement and sand and the face of the stack and the roof covering and rain penetrates these cracks. Due to frost the cracks open up and in a short time there is sufficient penetration of rain into the roof to cause serious damage.

Another method sometimes used is to bed tiles in cement and sand in the angle between the stack and the roof covering; the tiles are less likely to crack than the cement and sand fillet. But again, due to slight movements in the roof, cracks will open and rain penetrate. The only satisfactory way of permanently sealing the gap between a stack and the roof covering is to fix strips of sheet metal to the stack arranged so as to lap over other strips of sheet metal fixed in the roof covering. This at once accommodates any slight movements of the roof and seals the gap below. Any of the four metals in sheet form used as roof covering (see Vol. 1) may be used but lead or copper sheet are generally preferred because of their ductility.

Fig. 196 illustrates a view of a stack rising through the ridge of a roof showing lead sheet weatherings.

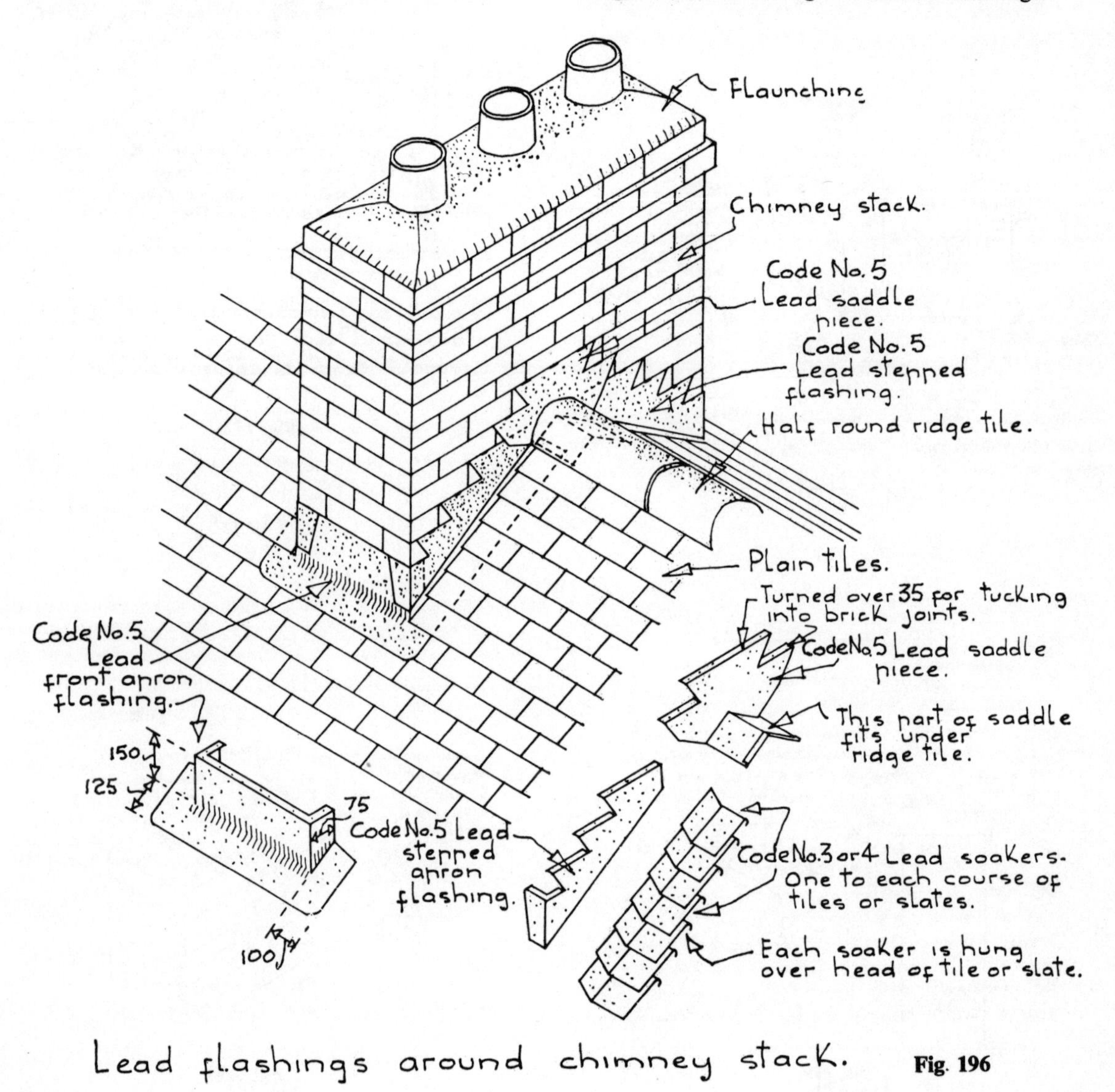


Lead flashings around chimney stack. **Fig. 196**

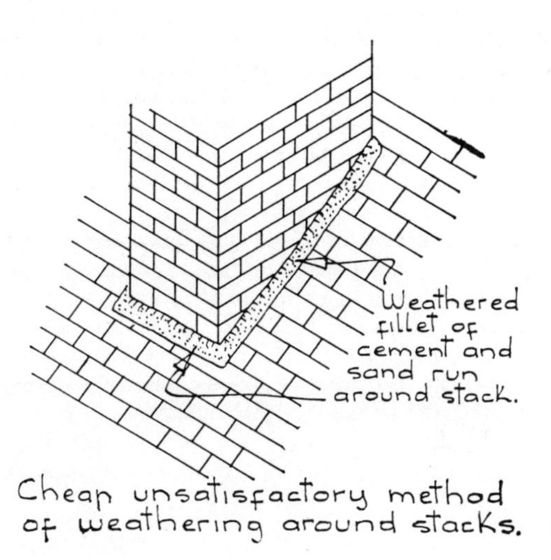


Fig. 197

**Soakers:** These are hung over the back of tiles or slates at the abutment of the roof covering and stack with part of the soaker dressed up against the wall as illustrated in Fig. 199. The soakers lap over each other down the slope of the roof so that water runs over them and on to the roof covering below the stack.

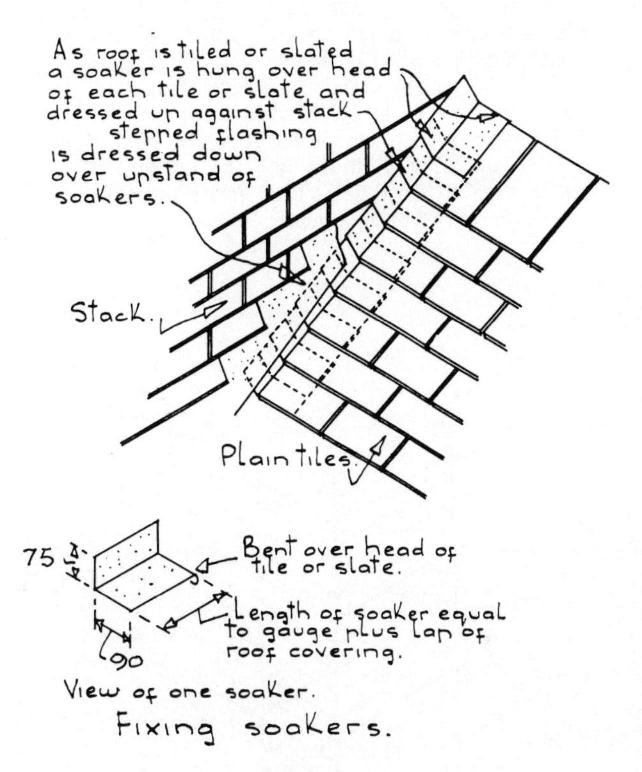


Fig. 199

**Stepped flashing:** A strip of sheet metal is cut so that its stepped edge can be fixed in horizontal joints in the side of the stack and the apron edge dressed down over the upstand of soakers. One method of setting out the cutting of this stepped flashing is illustrated in Fig. 198. The horizontal edges of the steps are tucked into raked out joints and wedged in position with wedges. Wedges are made by folding narrow strips of sheet until the wedge is sufficiently thick to wedge into the raked out joint.

70 degrees.
Brick courses
50
Brick courses
50
Water Line.
12
Edge of flashing.
Water Line.
Line of top of tile or slate roof covering.
Method of setting out sheet metal stepped apron flashing: Draw line of roof covering and brick courses to scale – Draw water line 65 above roof covering – Draw steps with slope of 70° to horizontal.

Fig. 198

**Saddle piece:** This is a piece of sheet metal cut with steps for wedging into joints and shaped to fit under the covering of ridge. Fig. 200 illustrates a saddle piece made from copper, zinc or aluminium.

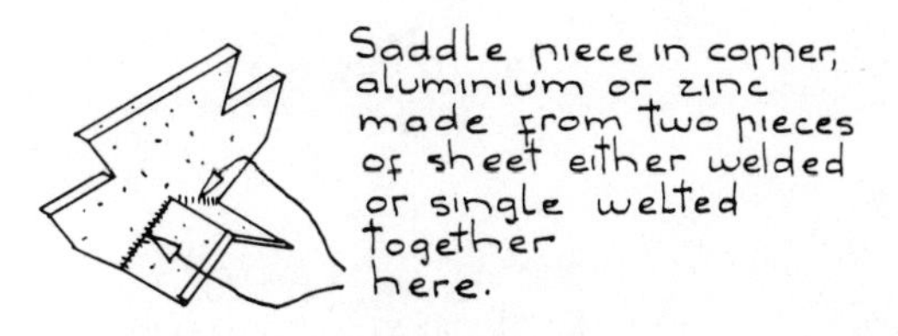

Saddle piece in copper,
aluminium or zinc
made from two pieces
of sheet either welded
or single welted
together
here.


Saddle piece in copper, aluminium or zinc sheet.

Fig. 200

**Front apron:** One sheet of metal cut and shaped to fit around the stack has its top edge tucked and wedged into a joint and its lower edge (apron) dressed over the roof covering. Fig. 203 illustrates an apron made from copper, zinc or aluminium sheet.

Where a chimney stack rises through a pitched roof below the ridge a sheet metal gutter has to be formed at the back of it as illustrated in Fig. 202. One piece of sheet metal is shaped to fit the back of the stack and over a small gutter formed at the back of the stack and under the roof covering. A separate apron is cut to fit the back of the stack so that it can be wedged into a joint and dressed down over the upstand of the gutter. Fig. 201 illustrates a back gutter and apron cut from copper, zinc or aluminium.

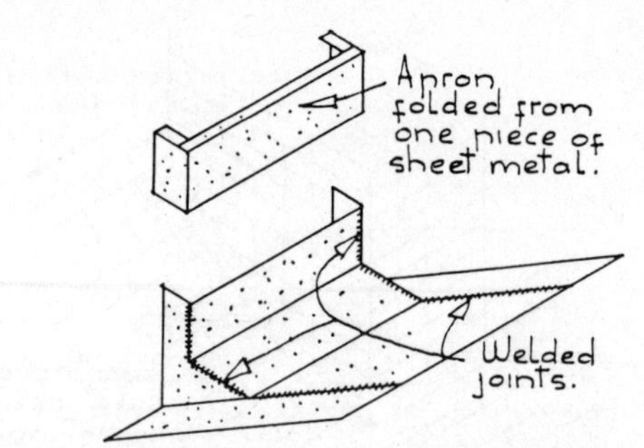


Fig. 201

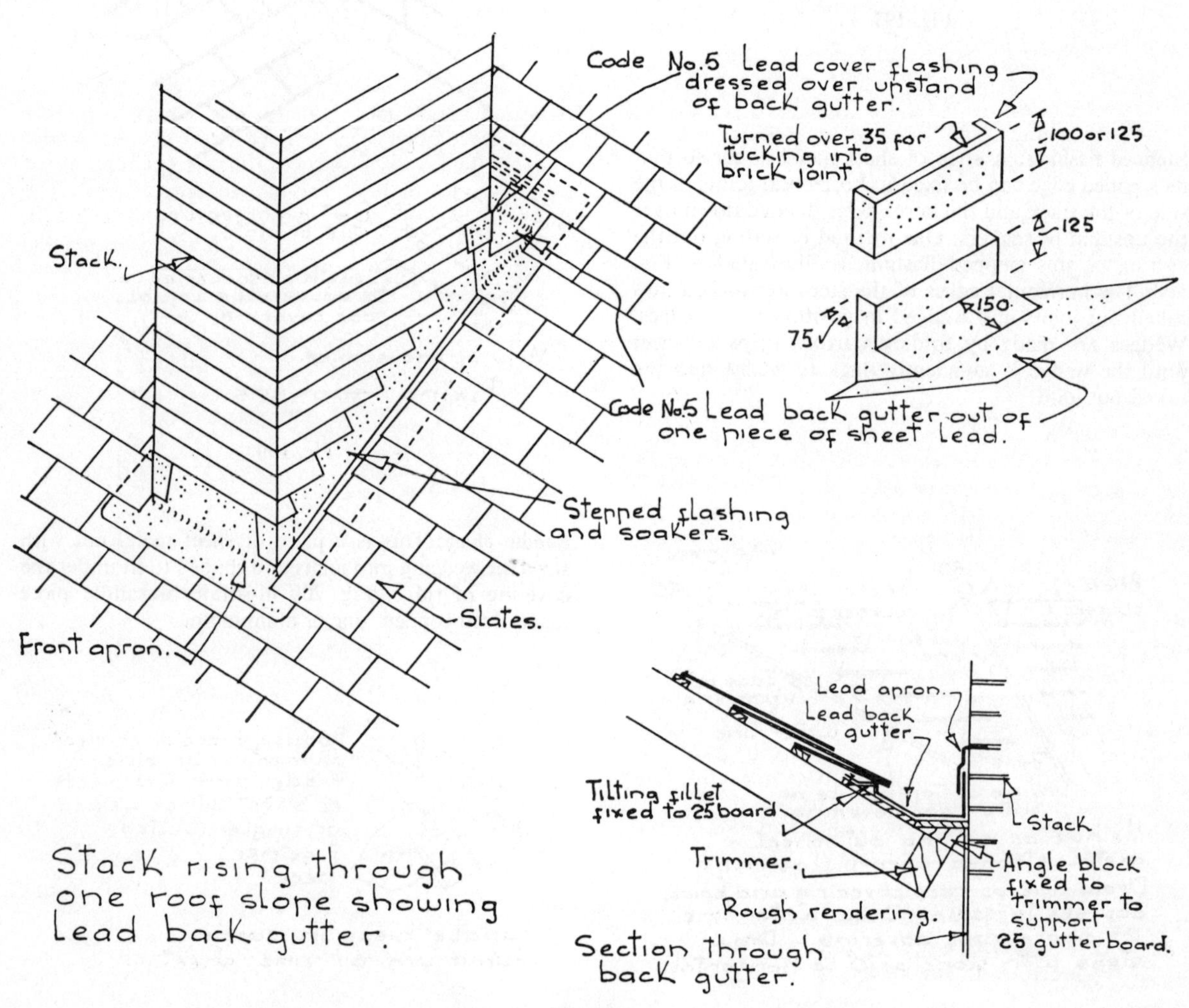


Fig. 202

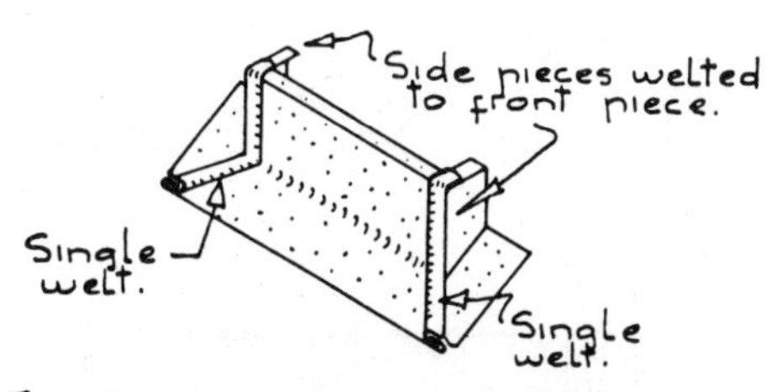


Front apron flashing in copper, aluminium or zinc sheet made from three pieces of sheet single welted together.

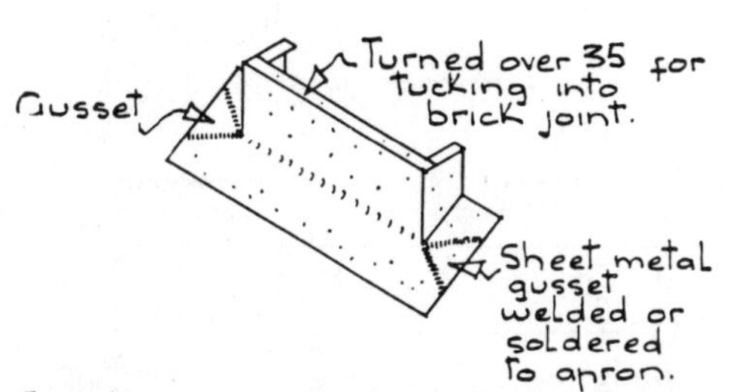


Front apron flashing in copper, aluminium or zinc sheet made by folding a sheet of metal and welding in gusset pieces [soldered in zinc].

**Fig. 203**

**Hearths:** The Building Regulations stipulate that every open fireplace should have a hearth of incombustible material extending under and in front of it. The hearth must extend at least 508 from the jambs, and at each side at least 152·4mm beyond the opening and be at least 152.4mm thick. Where the hearth is formed in a floor of combustible material (timber floor) its upper surface in front of the fireplace opening must not be lower than the floor surface. The minimum dimensions of hearths are illustrated in Fig. 204. The material most commonly used for hearths is concrete because it is cheap, strong and sufficiently resistant to damage by fire. Where the hearth is in a floor of monolithic reinforced concrete construction it is simply a part of the floor. If a timber floor is less than 150 thick the area of the hearth is raised above the floor so that it is at least 152·4mm thick.

With self-centering reinforced concrete floors and fire resisting concrete floors (Vol. 1) it is sufficient to fill the hollow blocks of concrete or terra-cotta, that are laid under the hearth, with concrete.

Where a hearth is formed in a timber floor the joists of the floor have to be cut and supported around it.

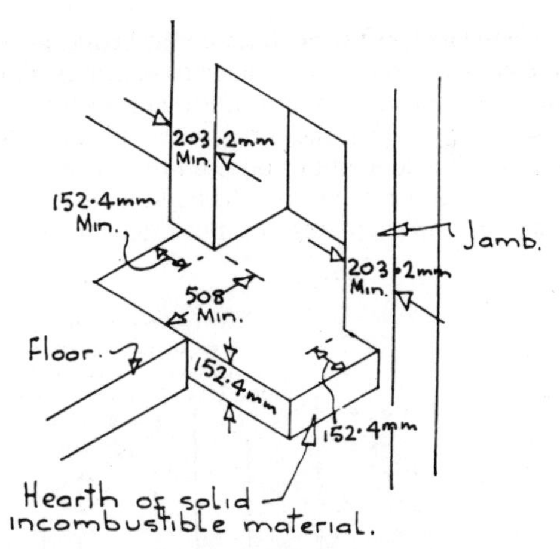


Diagram illustrating Minimum dimensions of hearths for open fires

**Fig. 204**

**Hearth in raised timber, ground floor:** The simplest way of constructing the hearth is to build a brick fender wall off the concrete oversite, fill the enclosed space with dry stone or brick and cast the hearth on it. The fender wall can be arranged to support both the hearth and the ends of the floor joists running towards the hearth, as illustrated in Fig. 205.

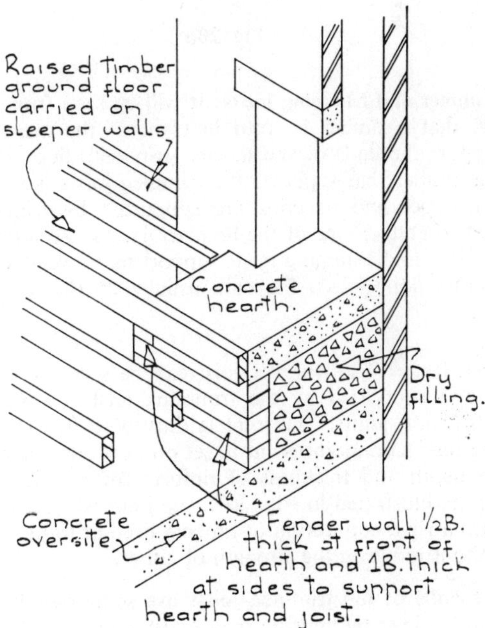


Hearth in raised timber ground floor.

**Fig. 205**

**Hearth in timber upper floor:** Because the hearth projects into the floor the timber joists have to be trimmed (cut) around it. But the ends of joists at the side of the hearth cannot be built into the brickwork in the jambs, because building regulations prohibit this. The relevant building regulations restricting the building in of timber around flues and fireplaces are illustrated in Fig. 206.

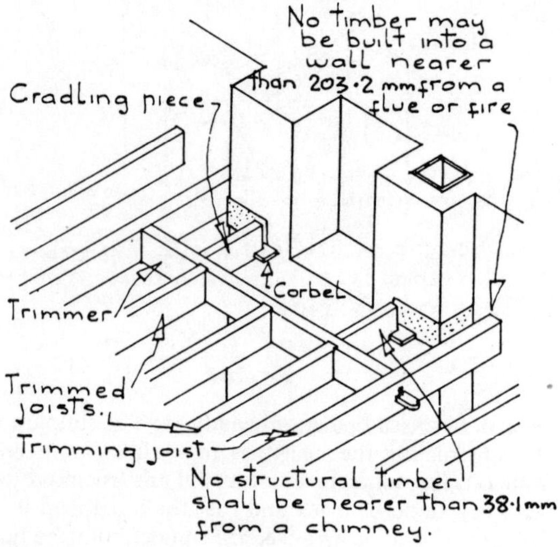


**Fig. 206**

**Trimmer and trimming joists:** It will be seen from Fig. 206 that a timber is fixed in front of the hearth to support the ends of the shorter (trimmed) floor joists. The timber that supports the trimmed joists is termed a trimmer and its ends are supported by trimming joists on either side of the fireplace jambs. Because the trimmer and trimming joists support more weight than normal floor joists, they are usually 25 thicker than them.

**Tusk tenon:** The trimmer joist can be securely joined to and supported by the trimming joists by making tusk tenon joints. The joint is illustrated in Fig. 207. The tusk tenon joint is cut to set proportions related to the depth and thickness of timbers from which it is cut, as illustrated in Fig. 207. The joint is designed to form a rigid connection of timbers fixed at right angles without reducing the strength of either.

The ends of the trimmed joists are supported by the trimmer joist by housing their ends in the trimmer for half their depth as illustrated in Fig. 208. Of the two joints illustrated the dovetail half depth housed is the better as the end of the trimmed joist is firmly anchored in its housing by the dovetail. If the floor joists of a

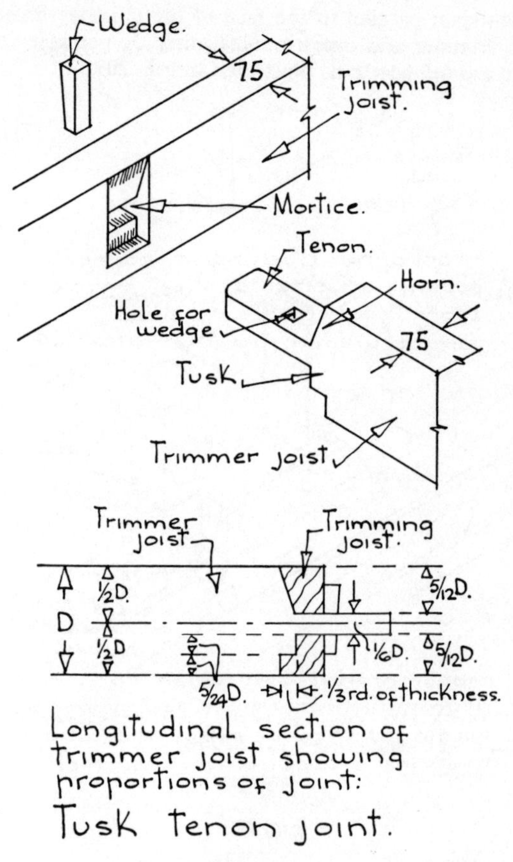


**Fig. 207**

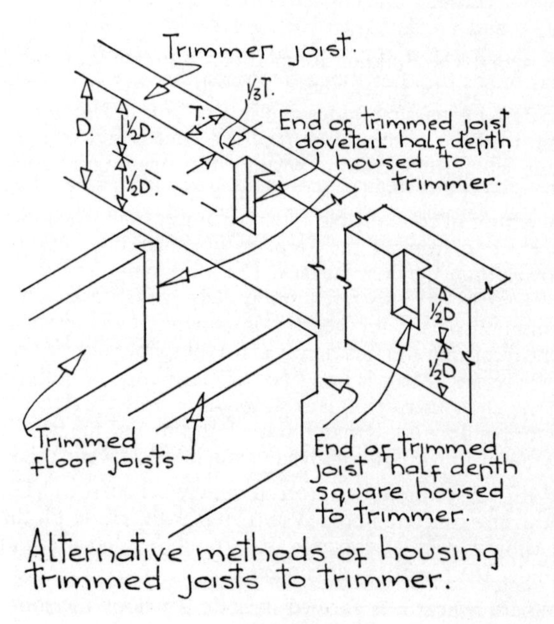


**Fig. 208**

room span parallel to the face of the chimney breast two trimmer and one trimming joists are necessary to trim around hearth as illustrated in Fig. 209.

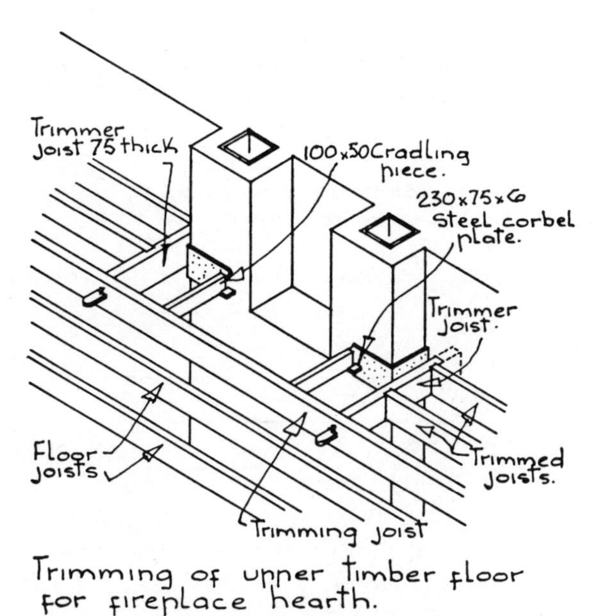


Fig. 209

**Method of supporting concrete hearth:** If the timber floor joists are deeper than the hearth thickness the timber around the hearth is used to support it, as illustrated in Fig. 210. If the hearth is as thick as the floor joists are deep it can be reinforced and built into jambs to cantilever out without support from floor joists as illustrated in Fig. 211.

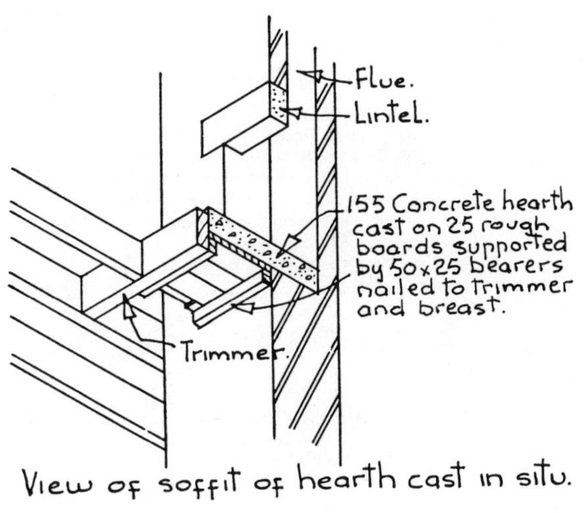


Fig. 210

The minimum distances permitted by building regulations between ends of structural timbers built into walls, and fireplaces or flues, are designed to prevent the possibility of the timbers being damaged by the heat of the fire. For the same reason the face of brickwork or blocks, less than 1B thick around flues, which pass through timber floors or roofs, has to be rendered with cement and sand (1 to 3) to a thickness of about 12. This coating of cement and sand is usually performed by the bricklayer as he builds the chimney and no great care is taken to leave the surface rendering smooth because it will be hidden in the floor or roof. Because it is left rough it is generally called rough rendering. No structural timber (i.e. floor joist) may be fixed nearer than 38·1mm from the face of the rough rendering described above.

**Cradling pieces:** Because the trimmer or trimming joists have to be built in at least 203·2mm from flues or fireplaces there is a space between them and the side of the concrete hearth. So to provide support for the ends of floor boards running up to the hearth, timber cradling pieces are fixed. One end of the cradling piece is nailed to the trimmer or trimming joist and the other borne on a brick or stone corbel projecting from the jambs as illustrated in Fig. 206.

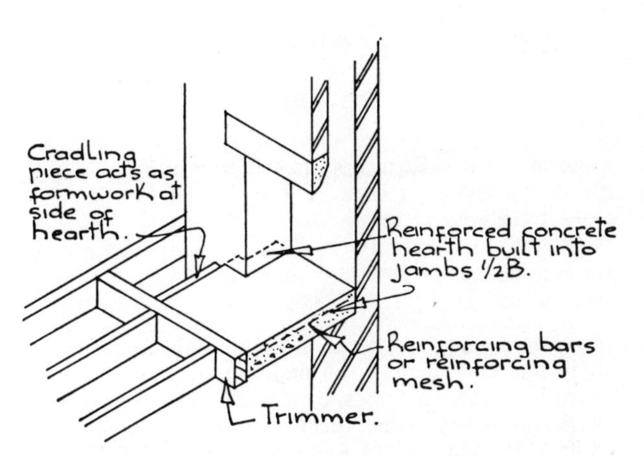


Fig. 211

**Trimming roof around stack:** No structural timber may be built in within a distance of 203·2mm from a flue. Timber roofs, therefore, have to be trimmed around stacks. The trimming of timber rafters and joists is illustrated in Fig. 212. It will be seen that the rafters and joists running into the stack are trimmed (cut), and their ends supported by trimmer joists. The ends of the trimmers being supported by trimming joists each side of the stack.

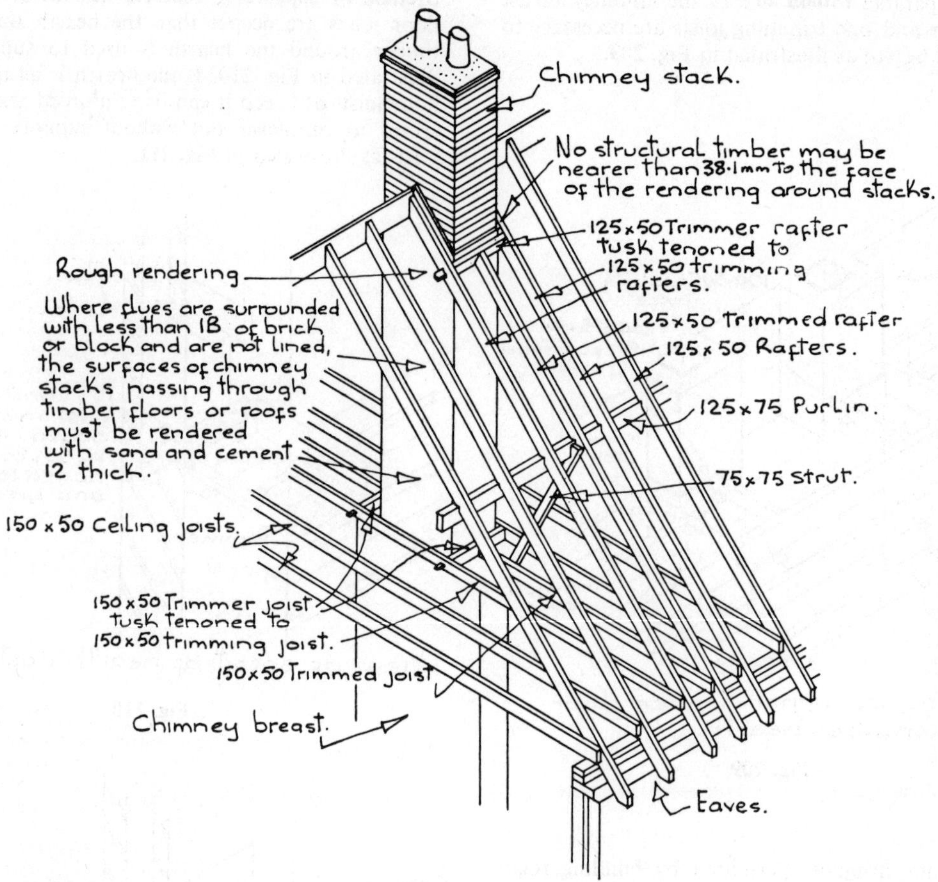


Fig. 212

**Fireplaces back to back:** It is economical to construct fireplaces back to back on opposite sides of internal partitions and on opposite sides of walls separating buildings (party walls) because there is a saving in brickwork, trimming of floors and roof and flashings as compared to the cost of building the fireplaces separately. Furthermore a fireplace built in an internal wall will not lose heat to cold external air as does a fire built in an external wall. Building regulations set out minimum thicknesses of brick or block behind fireplaces, as illustrated in Fig. 213.

**Fireplaces:** The size of a fireplace opening depends on the type of fuel to be burned and the size of the room that the fire heats. For rooms of up to 40 m$^3$ in size a 350 wide fire is recommended and for rooms up to 55 m$^3$ a 400 wide fire. Larger rooms should have means of heating in addition to open fires.

**Firebacks:** Fireplace openings for burning coal or coke should be lined with a fireback. Fig. 214 is an illustration of a typical one-piece fireback.

The purpose of the fireback is to contain the burning fuel, prevent the heat of the fire from damaging the wall behind it and to deflect some of the radiant heat out from the fire. The deflection of heat into the room is assisted by the splay wings of the fireback and the projecting knee.

Firebacks are made from fireclays which are clays that contain a high proportion of sand with alumina. The clay is moulded, dried and burned in a kiln. The burned fireclay is able to suffer considerable heat without damage and is used for this reason.

It will be seen from Fig. 214 that there is a projection in the back of the fireback and this is termed a knee.

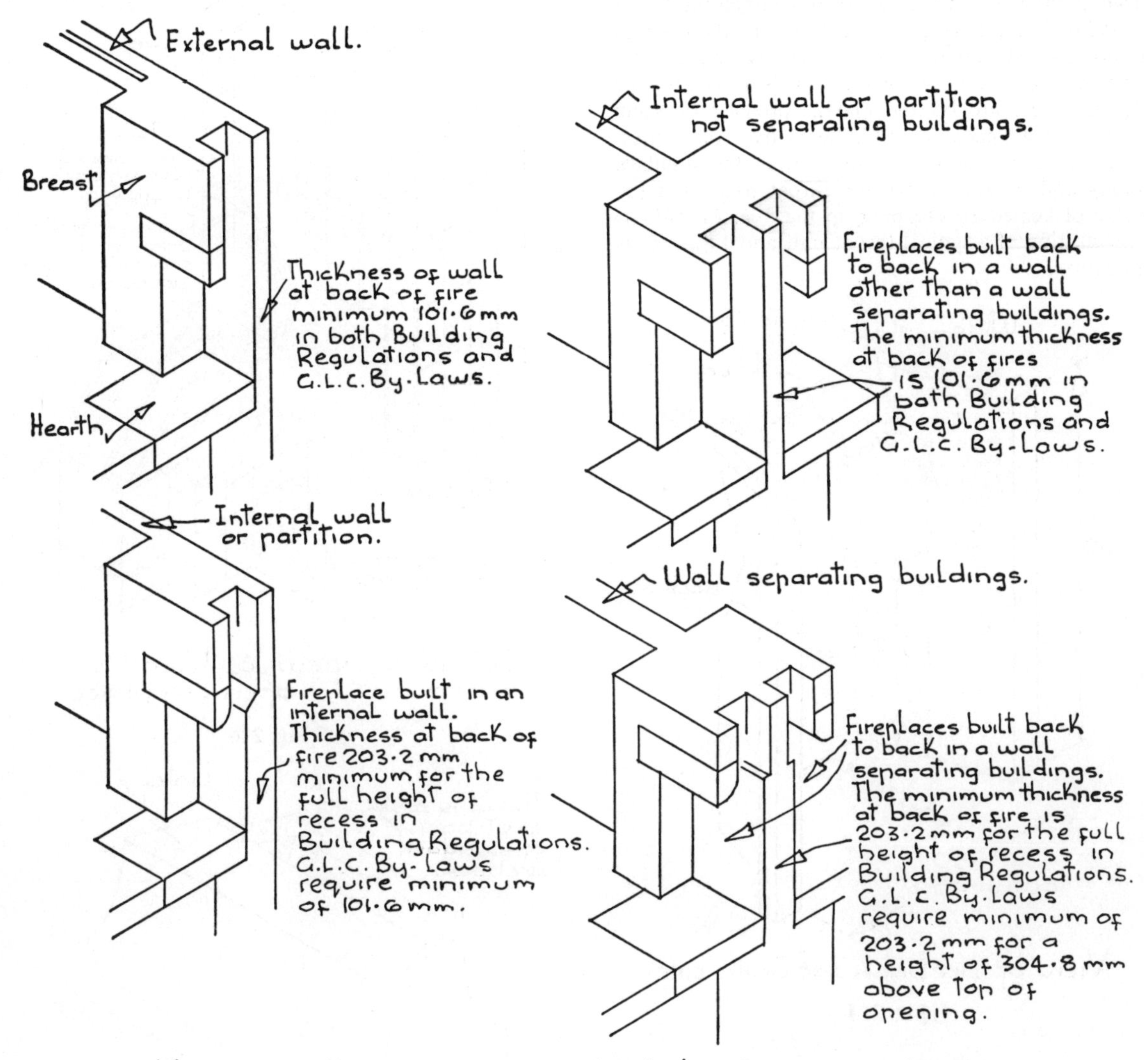

External wall.
Breast
Thickness of wall at back of fire minimum 101·6mm in both Building Regulations and G.L.C. By-Laws.
Hearth
Internal wall or partition not separating buildings.
Fireplaces built back to back in a wall other than a wall separating buildings. The minimum thickness at back of fires is 101·6mm in both Building Regulations and G.L.C. By-Laws.
Internal wall or partition.
Fireplace built in an internal wall. Thickness at back of fire 203·2 mm minimum for the full height of recess in Building Regulations. G.L.C. By-Laws require minimum of 101·6mm.
Wall separating buildings.
Fireplaces built back to back in a wall separating buildings. The minimum thickness at back of fire is 203·2mm for the full height of recess in Building Regulations. G.L.C. By-Laws require minimum of 203·2mm for a height of 304·8 mm above top of opening.

Minimum thickness of wall at back of fireplaces.

**Fig. 213**

The purpose of this knee is to reduce the amount of heat that rises up the flue by deflecting a part of it into the room, and also to deflect some radiant heat into the room. Firebacks are manufactured in either one, two, four or six pieces which can be put together. Fig. 215 illustrates typical sectional firebacks.

The advantage of firebacks made in two or more pieces is that they can be built into an existing fireplace when the old fireback has cracked and has been cut out, without having to take out the fireplace surround.

The width of the fireplace opening between brick jambs depends on the size of fireback to be built into it and is usually made the nearest half brick dimension wider than the fireback—e.g., for a 350 fireback an opening 2B wide; for an 450 fireback an opening 2½B wide.

A fireback is built into a fireplace opening with its front edges in line with the face of the breast so that the fire surround opening frames the fireback as illustrated in Fig. 216.

It will be seen from Fig. 216 that the space behind the fireback in the fireplace opening is filled with dry broken brick mixed with a little concrete to bind it together. The purpose of the brick filling is to act as an insulator against excessive loss of heat from the fire to the wall behind.

**Throat units:** Much of the heat from an open fire is lost up the flue because it heats air drawn into the fire and this hot air rises up the flue. If the opening directly above the fire is reduced some of this heat loss can be avoided. There are on the market pre-cast concrete throat units designed for building in over open fires. The pre-cast unit acts as a lintel over the fireplace opening and provides a narrow throat to restrict the volume of heated air escaping up the flue. Fig. 217 is an illustration of a typical throat unit and Fig. 218 its application.

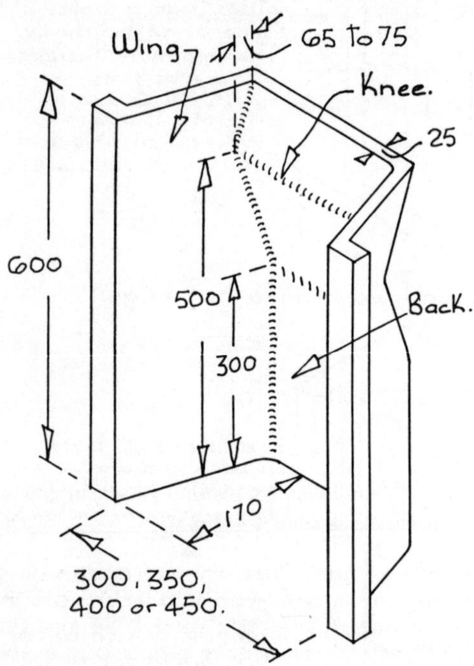


Fig. 214

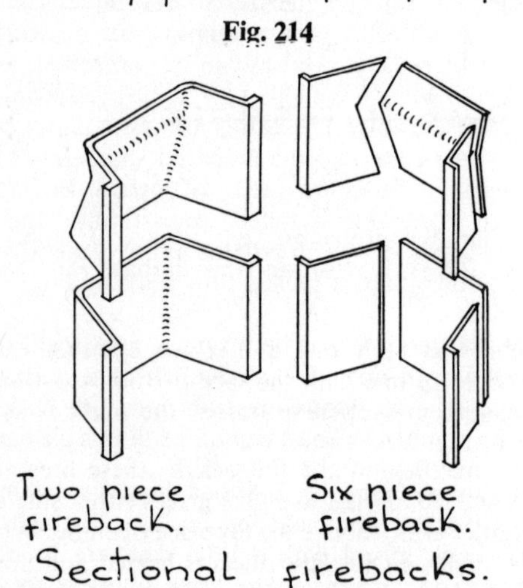


Fig. 215

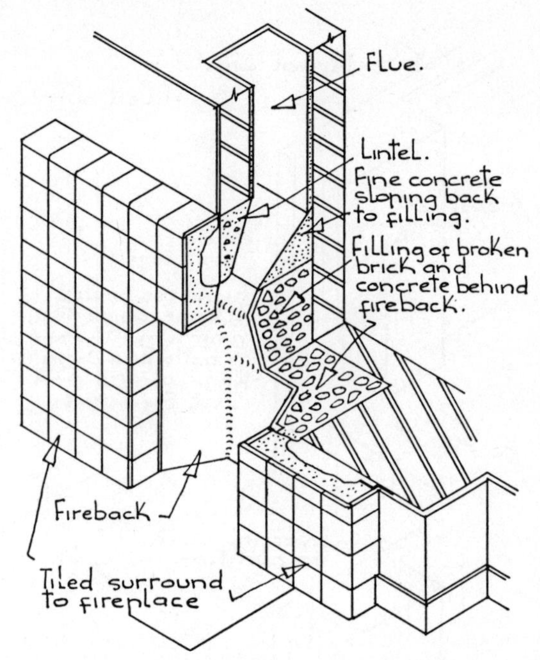


Fig. 216

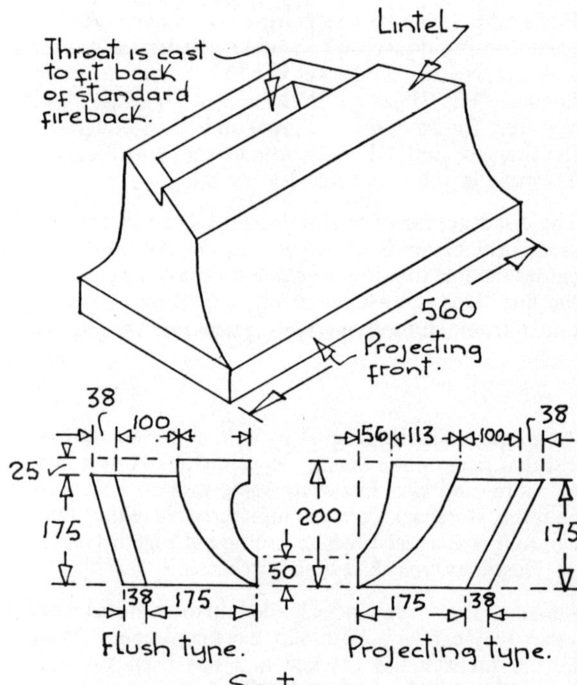


Fig. 217

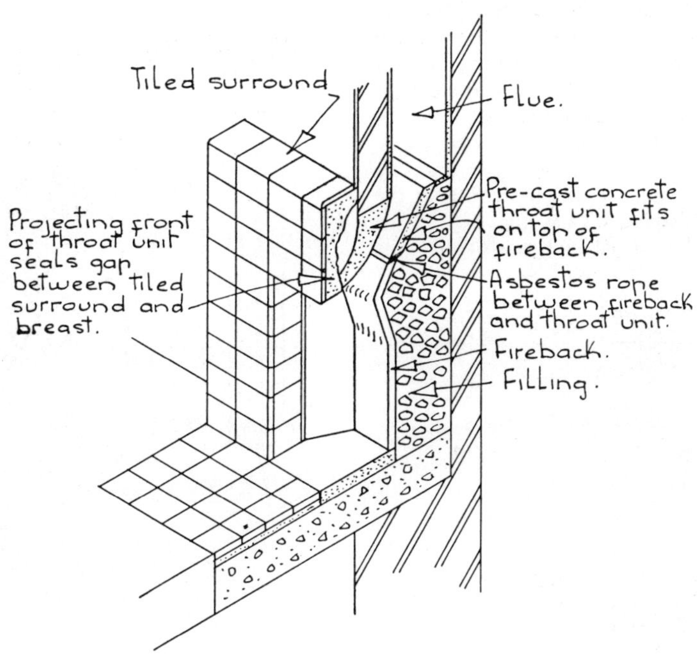


Fig. 218

**Fire grates—stool bottom grates:** The simplest form of grate consists of a cast iron basket on legs and a cast iron fret. Neither the basket nor the fret is fixed in the fireplace. Fig. 219 is an illustration of a typical basket and fret. The stool bottom grate and fret are placed in the fireplace and are made to suit standard firebacks. They can be taken out for clearing ashes.

The disadvantage of this grate and loose fret is that a great deal of air is drawn in around the fret which causes the fuel to burn rapidly and heated air passes up the flue. This is wasteful of heat and because the air drawn from the room has to be replaced, cold air from outside the room enters around doors and windows and causes objectionable draughts.

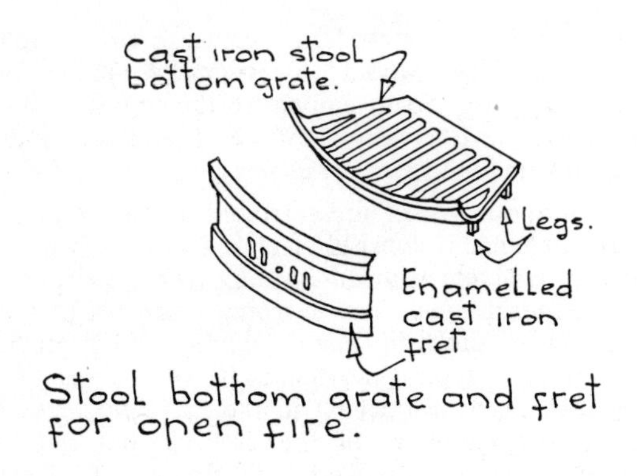


Fig. 219

**Slow burning fires:** These fires are made with grates that are fixed in the fireplace and the fret can then be locked to the grate so that there is an air-tight seal around the fret. The purpose of this is to enable the householder to control the amount of air drawn in through the slots in the fret to the underside of the grate. By controlling the amount of air drawn in, the rate at which the fuel burns can be controlled, and it is possible to keep a fire alight all night by slowing down its rate of burning. These fires are particularly suitable for burning coke and the prepared smokeless fuels on the market. As more areas of urban England are declared "smokeless zones" by statute, the slow-burning open fire will become more generally used. Fig. 220 illustrates a typical slow burning fire.

**Convector fires:** As has been said, a great deal of air is drawn into an open fire, and much of it is drawn up the flue. The modern convector open fire is designed to direct some of the heat contained in this air back into the room. Behind the fireback in these fires are cast iron ducts connected to intake grilles either side of frets and outlet grilles above the fireplace opening. When the fire is alight, air in these ducts is heated and flows out through the high level outlet grilles, and is replaced by cooler air drawn in which in turn is heated. Fig. 220 illustrates a typical convector fire.

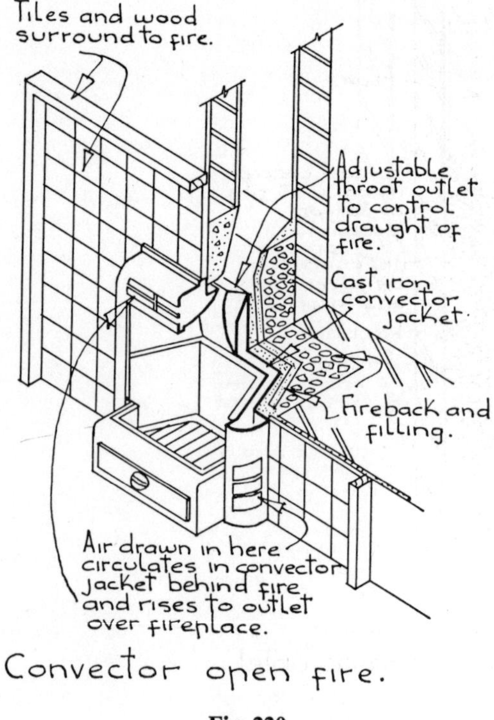


Fig. 220

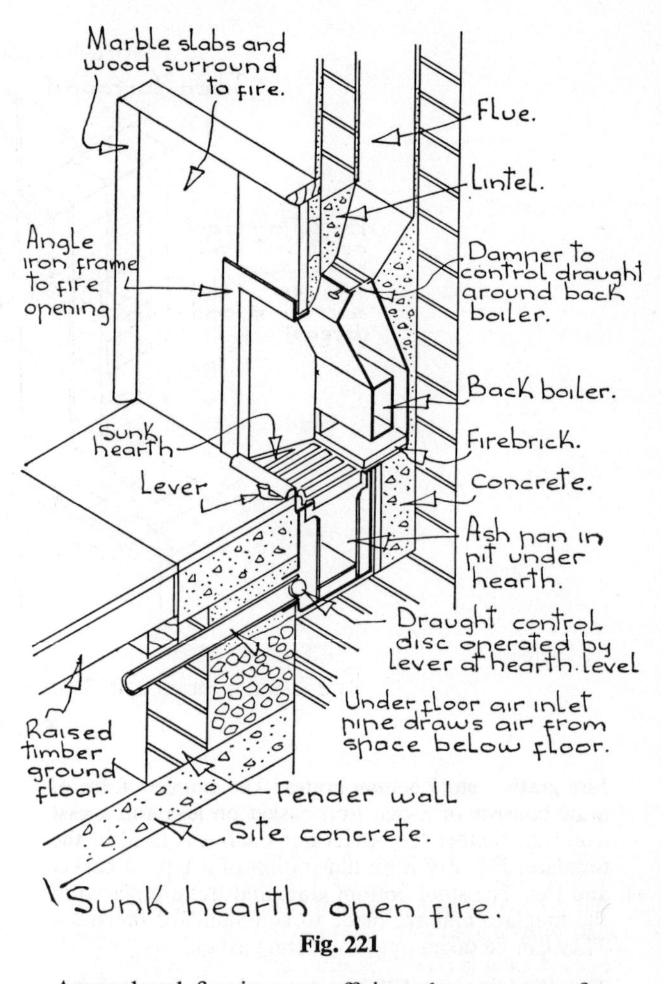


Fig. 221

**Sunk hearth open fire:** Both the ordinary open fire and the slow burning fire draw a considerable volume of air from the room and air entering the room to replace it causes draughts. The sunk hearth open fire is designed to reduce the volume of air drawn from the room and up the flue. This is achieved by constructing the hearth, inside the fireplace, at a level below that of the floor so that air can be conducted from outside to the sunk hearth. Fig. 221 illustrates the construction of this type of fire.

It will be seen that a pipe under the floor feeds air to the sunk hearth and the volume of air drawn through the pipe can be controlled by a flap at the mouth of the pipe. A deep ash tray is fitted in the sunk hearth, of sufficient capacity for the ash from several days burning. The sunk hearth fire can be made to burn quickly by opening the air inlet, or slowly—for overnight burning—by closing it. This type of fire greatly reduces the draughts experienced with other open fires.

**Enclosed grates and fires:** An open fire allows a great deal of heat to escape up the flue and causes draughts. Totally enclosed fires avoid much of the wastage of heat and do not cause so much draught. But they do not provide a visible cheerful glow and flame. Fig. 222 is an illustration of a typical enclosed fire for burning coal or coke.

An enclosed fire is more efficient than an open fire because combustion of the fuel is contained in it and little escapes up the flue and because the large heated surface of the stove spreads warmth around itself. The appearance of these stoves is not to everyone's taste, and the stove obstructs useful floor space.

**Fire surrounds:** The modern fireplace is comparatively small and by itself would look insignificant in any room of average size. So to emphasise the fire and also to cover the joint between the fireback and the jambs a fire surround is used.

The cheapest type of fire surround is known as a tile slab surround. It consists of glazed tiles fixed on to a core of concrete with an opening in the surround to suit standard fires. The surrounds are pre-cast and delivered to site ready for fixing. Fig. 216 is an illustration of a typical surround. It will be seen from Fig. 216 that the opening in the surround laps over the projecting wings of the fireback and the joint between the two is filled with fire-resisting cement (High Alumina cement).

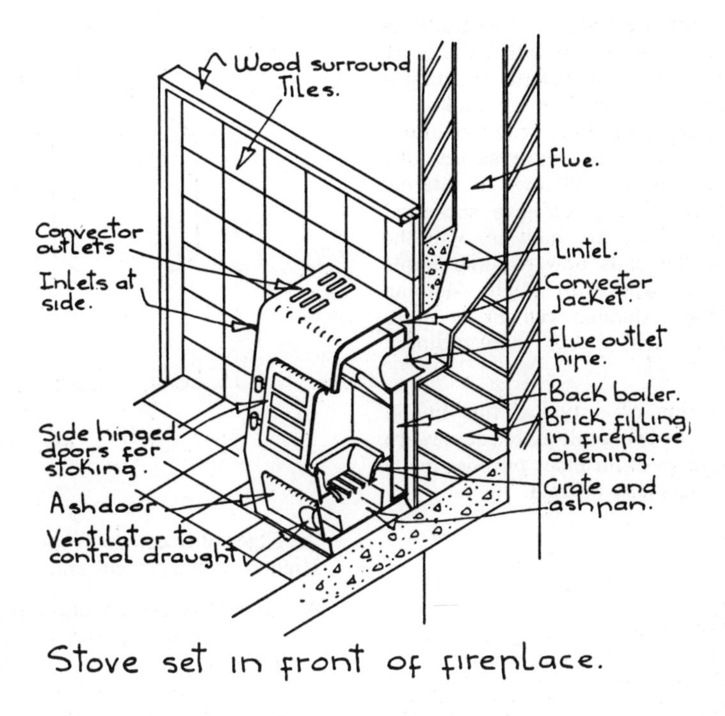


**Fig. 222**

The surround is screwed to the jambs of the fireplace through steel clips cast into the back of the surround and projecting at the sides. These clips are screwed to lead plugs driven into holes in the jambs.

Instead of a pre-cast tile slab surround, tiles may be fixed directly to the brickwork around the fire and enclosed in a wood or metal frame. The joint between the tiles and the fireback is usually masked with a metal frame made for the purpose and secured in position with lugs welded to the back of it. This latter frame is usually made from mild steel angle with a facing of stainless steel, bronze or chromium plate strip.

Slabs of a natural decorative stone may be used instead of tiles. The stone is bedded in mortar on the brick-work around the fire and finished with a metal frame to the fireback, as illustrated in Fig. 221.

Any material which is considered decorative and which will not be damaged by the heat of the fire may be used as a fire surround.

Various arrangements of brick, natural stone, sheet metal, toughened glass and cast iron are used with or without wood and metal frames.

**Flue liners:** The conventional flue is lined with par-getting which is roughly spread on the inside of the flue as it is built. The surface of this pargetting is not particularly smooth and does to some extent impede the flow of smoke up the flue.

With ordinary open fires this is of no consequence as the great volume of air drawn into the fire generally causes the smoke from it to flow rapidly up the flue. But with slow burning open fires and boilers the restriction of smoke, due to rough surfaced pargetting, may seriously reduce their efficiency and their per-formance can be improved if the flue is lined with some smooth-surfaced material.

The smoke caused by burning such fuels as coal, coke and anthracite contains water vapour, sulphur com-pounds and tar acids. The smoke from slow burning open fires rises slowly in flues and cools sufficiently for its water vapour to condense on the flue walls. The drops of water containing dissolved sulphur com-pounds and tar products will saturate the flue walls and will attack lime or Portland cement in the parget-ting and in the mortar between bricks, blocks or stone. The resultant water-soluble salts in the condensed water may crystallise and cause considerable expansion and cracking in pargetting and mortar. The expan-sion may cause cracking of the external faces of the flue or staining and cracking of plaster surfaces internally. Obviously this effect will be more pronounced in flues on external walls as cold outside air will cause con-siderable condensation in them.

It is considered good practice today to line the flues of slow burning open fires and boilers with some dense smooth material such as glazed stoneware or asbestos cement which will not restrict the flow of smoke and will be impervious to attack by sulphur or tar solutions. Salt glazed stoneware drain pipes have successfully been used to line flues. Either 100 or 150 internal diameter pipes are used. The pipes have socketted ends and are 600 long. They are built in with the chimney with their socket ends upwards and joints between the pipes are made with High Alumina cement and sand. High Alumina cement is more resistant to sulphate or acid attack than ordinary Portland cement.

Fig. 223 is an illustration of a typical flue lining. The flue is gathered over by using salt-glazed stoneware bends. As an alternative to drain pipes purpose-made fireclay or terra-cotta liners may be used.

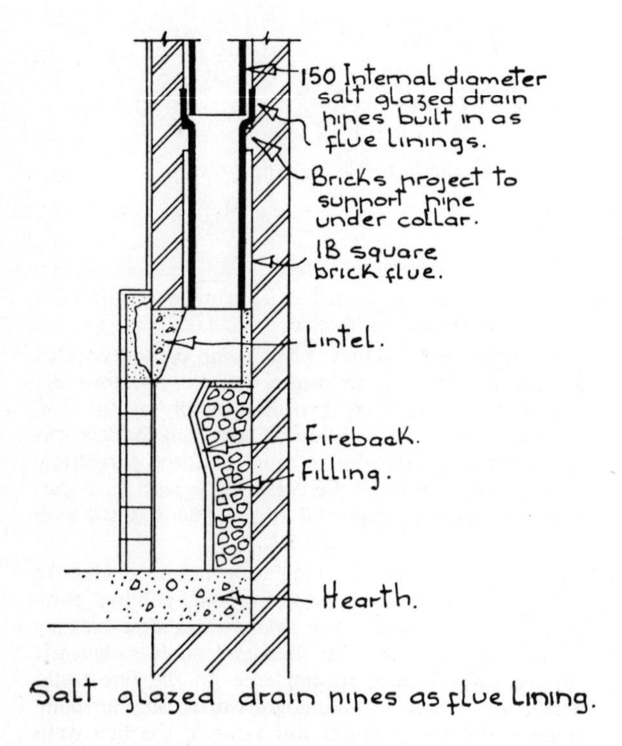


**Fig. 223**

Fireclay or terra-cotta flue liners are made in lengths of 290, to suit brick course heights, and either square with rounded corners or round in section, as illustrated in Fig. 224.

These liners are built into chimneys and supported by bearing bars built across the angle of flues and under

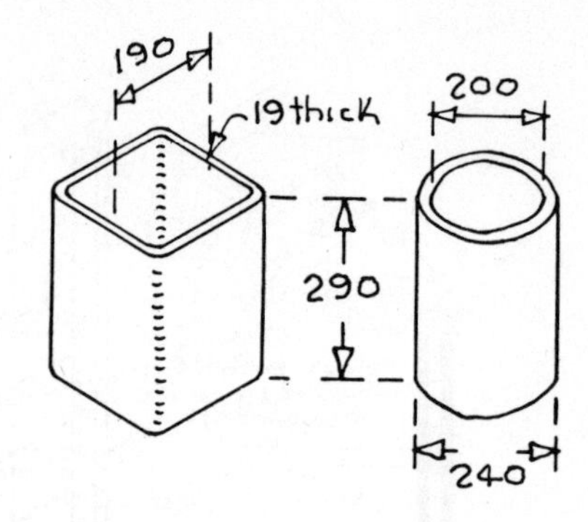


**Fig. 224**

the edges of liners. Joints between liners are made with High Alumina cement and sand.

As an alternative to drain pipes or clay flue liners asbestos cement flue pipes may be used. These pipes are made in lengths of 600, 900, 1·2 or 1·8 and an internal diameter of 100 or 150 is satisfactory for most slow-burning fires and domestic boilers. The pipes have socketted ends and are built in the same way as drain pipes. Of the three types of flue liner noted, salt glazed stoneware drain pipes of good quality are the most resistant to damage caused by smoke condensing in flues.

Obviously a flue lined with one of the liners noted is more expensive than one which is pargetted. But the small additional expenditure is well worth while for slow burning fires and domestic boilers, particularly when their flues are in an external wall.

**Pre-cast concrete flue blocks:** Domestic gas fires require a smaller flue than open fires burning solid fuels. A range of pre-cast concrete blocks, with flueways in them of sufficient size for gas fires, is manufactured. The flue blocks are made to bond in with brickwork courses so that the blocks can be built into brick walls or partitions without a projecting breast, as illustrated in Fig. 225. A variety of straight, offset, raking and stack blocks is made as illustrated. Building in blocks are made so that the gas fire can be fixed in a recess in the wall.

The blocks are manufactured from Portland cement, Portland blast-furnace cement or High Alumina cement with a natural aggregate such as gravel. The blocks are laid in lime mortar mix, 1–3 of lime and cement.

The flueway for gas fires of up to 4·5 kW is 300 long, and that for fires of over 4·5 kW 375 and both are either 50 or 60 wide. The letters kW are an abbreviation of kilo Watt which is used as a measure of the heat output of fires and boilers.

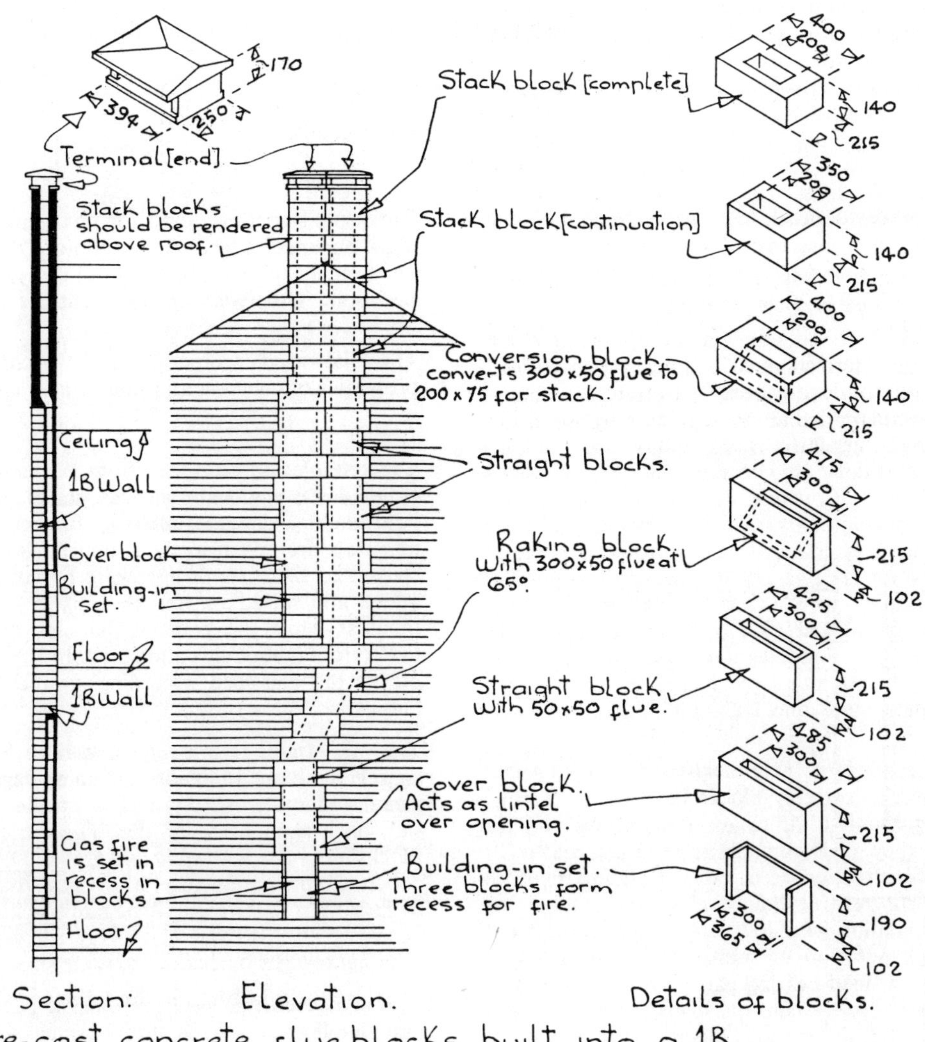


**Fig. 225**

### Reference Books and Publications:

**British Standards:**

No. 1181. Clay flue linings and chimney pots.
No. 1251. Open fireplace components.
No. 1281. Glazed ceramic tiles for walls and fireplaces.
No. 1289. Pre-cast concrete flue blocks for gas fires and ventilation.
No. 2845. Coke-burning inset open fires.
No. 3128. Inset open fires.
No. 3376. Open convection fires.

**British Standard Code of Practice:**

CP. 131.101. Flues for domestic appliances burning solid fuel.
CP. 403. Open fires, heating stoves and cookers burning solid fuel.

**Building Research Station Digests:**

No. 18. Smoky chimneys. (First Series).
No. 35. Heat loss from dwellings. (First Series).
No. 60. Chimney design for domestic boilers.

## CHAPTER SIX

# PARTITIONS

### SfB (22) Partitions: General

Partitions serve to divide the space inside buildings into rooms.

If a partition aids in supporting floors and roofs it is described as a load bearing partition.

Partitions which in no way aid in supporting floors and roofs are termed non-load-bearing partitions. Obviously a non-load-bearing partition can be of lighter construction than a load-bearing partition. The materials generally used today for building permanent partitions are concrete blocks, clay blocks, brick or timber. Concrete or clay blocks are the materials most commonly used. The advantage of these blocks is that they are mass produced in a range of thicknesses from 51 up to 219, from which it is possible to select one sufficiently thick to be stable when used as a non-load-bearing partition or thich enough to be stable and support loads as a load-bearing partition. Because the blocks are large compared to bricks a partition can be more quickly built with them than with bricks.

Another advantage of these blocks is that lightweight concrete and hollow clay blocks have better thermal insulation per unit of thickness than bricks and for this reason they are commonly used in external walls of cavity construction (see Vol. 1). Being light in weight the concrete or hollow clay blocks require a slighter foundation and less strengthening of floors than a brick partition of similar thickness. Brick is used mostly for load-bearing partitions in buildings of two or more storeys where the loads from floors and roof are such that either a solid block partition, say 102 thick, or brick has to be used. A half-brick thick partition of common bricks is generally slightly cheaper than a similar partition built with 102 thick dense aggregate concrete blocks.

Up to the end of the nineteenth century partitions were commonly built of timber. It was then a comparatively cheap building material.

Today a timber partition plastered both sides is more expensive than a partition of similar thickness built of blocks plastered both sides. A timber partition is a less efficient sound insulator than a block partition of the same thickness and if it is used as a load-bearing partition it has to be protected from damage by fire which adds to the cost of it.

The one advantage of a timber partition is that it is lighter in weight than either hollow clay or lightweight concrete block partitions of similar thickness.

Timber is sometimes used for non-load-bearing partitions on upper timber floors because the floor has to be strengthened less than it would for a similar block partition. Otherwise timber partitions are not much used.

The following is a description of the composition and properties of concrete and hollow clay blocks.

**Concrete blocks:** are made from Portland cement and either dense aggregate such as gravel and sand or a lightweight aggregate such as furnace clinker. The materials are mixed in the proportions of one part of cement to six parts of aggregate by volume. Water is added and mixed in, and the mix is consolidated in moulds. After the concrete has set the blocks are taken from the moulds and cured. In B.S.2028 concrete blocks are classified according to their composition and use as follows:

*Type A—Dense Aggregate Concrete Blocks:* Dense aggregate blocks are made with an aggregate of natural clean sand and crushed gravel, or crushed blast furnace slag or broken brick or tile. The majority of these blocks are made with sand and gravel. Dense aggregate concrete blocks are heavier and less resistant to transfer of heat than lightweight aggregate blocks but they are generally cheaper and can safely carry greater loads than lightweight blocks of the same size.

They are manufactured in the following sizes:

Length 448.

Height 143 or 219 to bond with bricks 65 deep.

Thickness 51, 64, 76, 102, 152 or 219 (hollow).

The 51 and 64 thick blocks are suitable only for non-load-bearing partitions. The other thicknesses can be used for the skins of cavity walls or for load-bearing partitions. The 219 blocks are cast with cavities in them as illustrated in Fig. 226.

The purpose of the cavities is to reduce the weight of the blocks to facilitate handling and laying. The manufacturers describe these blocks as having integral cavities implying that the cavities in the blocks serve the same purpose as the continuous cavity in a cavity wall. This is misleading as it is apparent that the cavities do not prevent rain soaking through the concrete surrounding the cavities in these blocks. These blocks are not superior to a solid brick wall 1B thick or a wall of solid concrete blocks 219 thick in

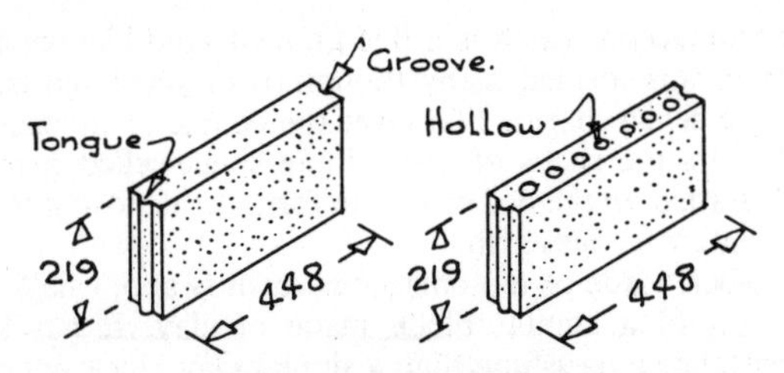

Solid and hollow concrete blocks.

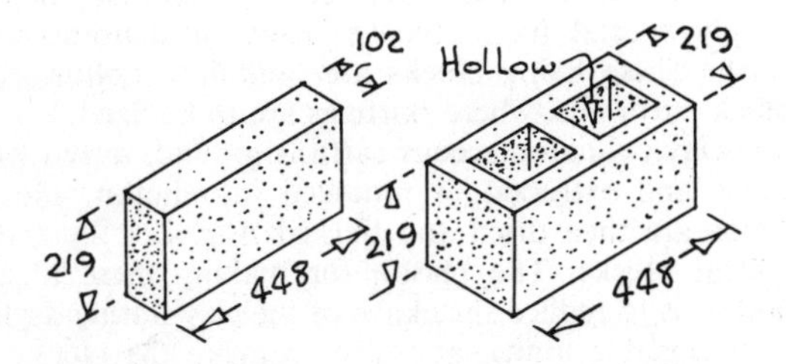

Solid and hollow concrete blocks

Typical concrete blocks.

**Fig. 226**

preventing rain soaking through to the inside face of the wall or in resisting transference of heat.

The 219 thick hollow dense aggregate blocks are used principally for external walls of bungalows and two

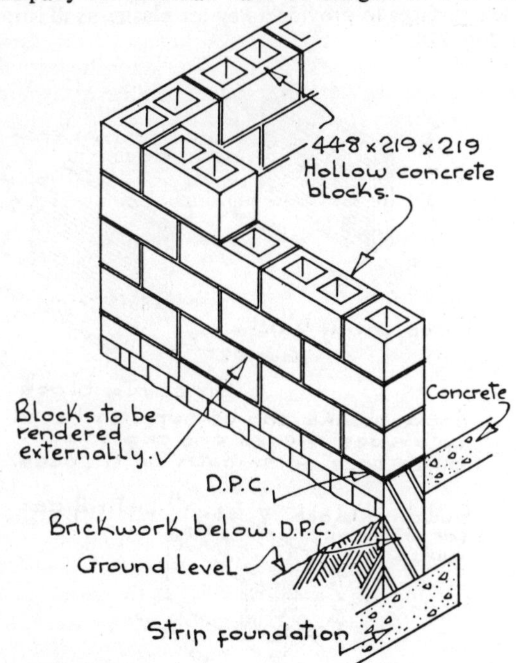

External wall of hollow concrete blocks.

**Fig. 227**

storey houses. They are laid in stretcher bond in cement mortar as illustrated in Fig. 227.

**Lightweight aggregate concrete blocks:**
*Type B for load-bearing walls and load-bearing partitions:*
The blocks are made from Portland cement and one of the following aggregates:

Granulated blast furnace slag.
Foamed blast furnace slag.
Well burned furnace clinker.
Expanded clay or shale.
Natural pumice.
Fly ash.

**Blast furnace slag:** Iron ore is smelted in blast furnaces to separate the iron from the other elements in the ore. Molten iron collects in the base of the furnace and earthy impurities from the ore combine with lime to form a slag which melts and floats on top of the molten iron. This molten slag is tapped off. If the molten slag is allowed to cool and solidify in air it becomes a dense, fine grained, sand-like mass. The slag is crushed and is used as an aggregate for concrete blocks.
If the molten slag is treated with water the resultant steam gives it a cellular structure and the slag then solidifies to form a porous, light-weight material known as foamed blast-furnace slag.

**Well-burned furnace clinker:** The residue from furnaces burning coke at high temperatures is an irregular shaped stone-like mass known as clinker. If the clinker has been thoroughly burned to remove practically all bitumen, it is suitable for use as a lightweight aggregate. The clinker is crushed for use as aggregate. It is cellular and lightweight and is the aggregate most commonly used for lightweight blocks.

**Expanded clay and shale:** Some clays and shales when roasted expand into a reasonably hard, cellular, lightweight mass suitable for use in building blocks. The expanded clay or shale is crushed for use as an aggregate.

**Natural pumice:** This is a cellular mineral, of volcanic origin, imported into this country. It is expensive compared to the other materials and is not much used today.

**Fly ash:** Is the very fine ash residue produced by steam turbines used to generate electricity. The quantity of this ash produced is an embarrasment to the Electricity Boards and they sell it on very favourable terms. The ash is treated with water and heated to form small pellets which are porous and light-weight.
Because the aggregates used in these blocks are cellular, the blocks are at once comparatively light in weight and better thermal insulators than dense aggregate blocks, bricks, concrete or stone.
They are lightweight and can advantageously be used for load-bearing partitions, requiring a slighter foundation than a partition of dense aggregate blocks.

Lightweight blocks are better thermal insulators than dense concrete, brick or stone, and are sufficiently strong to carry the load of a floor and a roof, and have largely been used for the inner skin, or for both inner and outer skins of cavity walls for houses and bungalows, and for panel walls in larger steel and concrete framed buildings. When used in the external face of buildings these blocks are covered with rendering, tile or slate as they readily absorb moisture and do not have an attractive colour or texture. They are made in the following sizes:

Length 448

Height 219 to bond with 65 deep bricks.

Thickness 76, 89, 102, 152, 203 or 219.

**Lightweight aggregate concrete blocks.**

*Type C—For non-load-bearing partitions:* These are made from Portland cement and any one of the lightweight aggregates previously described. They are made in the same length and height as those of Type B, but in thicknesses only of 51 and 64. These blocks are suitable only for non-load-bearing partitions. Their principal advantage is that being lightweight they require less strengthening of floors to support them than similar dense aggregate blocks.

**Breeze blocks:** Are made from lightly sintered (fused) gas coke breeze. The word breeze is used to describe the fine dust of gas coke which is too small for use as fuel. This coke dust is sintered (fused) into larger particles and used as an aggregate for concrete blocks. It is sold by the Gas Boards at a low price so that the blocks made from it are considerably cheaper than other concrete blocks. It is mixed with Portland cement and water and cast into blocks.

Coke breeze blocks are one of the clinker aggregate lightweight concrete blocks Type B or C and are made in the standard sizes noted. The coke sometimes used in the manufacture of these blocks is not as well-burned as it might be. If the coke breeze contains too much combustible matter the blocks will suffer considerable moisture movement which will cause cracking of plaster and renderings applied to their surface. In general it is wise to avoid using coke breeze blocks in the external face of walls or in any position where they are likely to absorb moisture and then dry out.

**Hollow clay blocks, B.S.1190:** Are manufactured from clays or diatomaceous earth. The clays used are those suitable for use in the manufacture of semi-engineering or engineering bricks, for example gault clay or wealden clay. This latter clay is one of the clays used in the manufacture of terra-cotta. The words "terra-cotta" mean earth burned and they could be used to describe any burned clay product. But by usage the words "terra-cotta" describe burned clay which is dense, hard, semi-vitreous and does not readily absorb water.

Diatomaceous earth is a fine grained sand-like deposit which was formed many thousands of years ago from single celled marine life. Over the course of thousands of years the shells of this minute single-celled marine life settled on the bed of seas to form the deposit known as diatomaceous earth.

A block made of diatomaceous earth is about half the weight of a similar block made of clay. It has less resistance to crushing than a similar clay block but can more readily be cut and nails and screws can easily be driven into it. The manufacturers of hollow clay blocks supply special fixing blocks made of diatomaceous earth. These fixing blocks are built into hollow clay block partitions, where skirtings are to be fixed.

The clay or diatomaceous earth is ground, mixed with water and extruded or moulded into hollow blocks which are then dried and burned. Fig. 228 illustrates typical blocks. The reason for making these blocks hollow is to reduce shrinkage of the clay during drying to reasonable limits, and also to make the blocks as lightweight as possible.

Hollow clay blocks do not suffer moisture movement due to changes in their moisture content as do most lightweight aggregate concrete blocks and are therefore unlikely to be the cause of cracking in plaster surfaces. Hollow clay blocks are highly resistant to damage by fire, and are good thermal insulators.

The finished surface of these blocks is so dense and non-absorbent that plaster would not adhere to it. The blocks are therefore made with shallow indentations in their surfaces to provide a key for plaster as illustrated in Fig. 228.

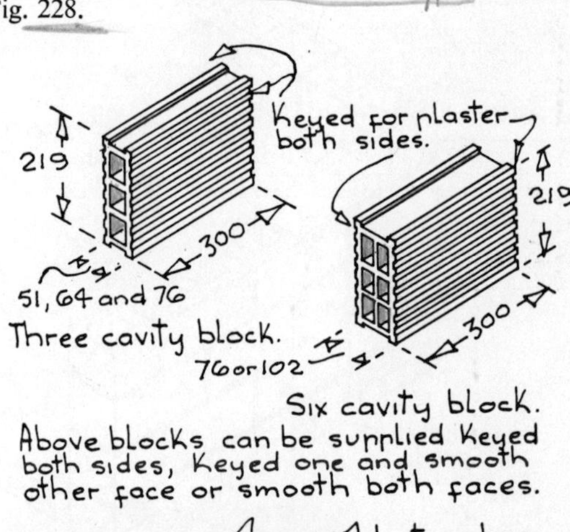


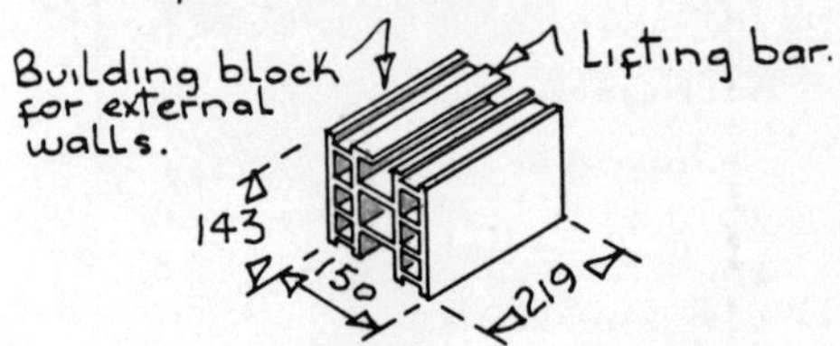


**Fig. 228**

Hollow clay blocks are made with both faces keyed for plaster or with one face keyed and the other smooth, or with both faces smooth. Smooth faced blocks are supplied for use where they will be painted or left exposed as in farm or factory buildings.

Bond blocks of the same height and thickness as full blocks, but either three quarters or half length are made and supplied.

**Thickness of block partitions:** The thickness of block required to build a stable partition depends on the size of the partition and, if it is load-bearing, the loads it supports. If a large partition, say 3·0 high between floor and ceiling, were built of 51 thick blocks, vibration caused by doors slamming or funriture being pushed against it would tend to overturn the blocks. This would cause plaster to crack or with heavy vibration the partition might collapse.

The following minimum thicknesses of blocks for partitions is set out in the British Standard Code of Practice C.P.121 :

Height between floor and ceiling or unrestrained width as between walls, partitions and storey-height door frames, whichever is less:

| *Height or length* | *Thickness* |
|---|---|
| Up to 2·4 | 51 |
| 3·0 | 64 |
| 3·6 | 76 |
| 4·5 | 102 |
| 6·0 | 152 |
| 7·5 | 219 |

Load-bearing partitions should generally be at least 76 thick which is the thickness required to aid in supporting one timber domestic floor or a timber roof. A partition 102 thick is generally sufficient to aid in supporting one domestic timber floor and roof.

**Load-bearing partitions foundation:** The loads on the base of load-bearing partitions in small buildings such as two-storey houses are usually much less than the loads on the base of the external walls, because partitions can be less thick than walls which have to resist rain penetration. For example the load on the foundation of a two-storey brick wall is generally about 30 to 45 KN/m of wall, whereas the load on the foundation of a 102 thick load-bearing partition built of lightweight blocks is generally about 15 KN/m of partition.

In Vol. 1, Chapter 1 it was explained that a foundation for a wall must be sufficiently wide to spread the loads on it to an area of subsoil capable of safely carrying the loads and also that the foundation must be sufficiently deep to avoid damage from differential movements which occur in some subsoils. The loads on the base of a load-bearing partition must likewise be spread to an area of subsoil capable of safely supporting them. The foundation of a partition, being within the protecting external walls, is not liable to damage from frost heave or volume changes in clay subsoils and therefore need not be carried so far down below the surface as an external wall foundation. The usual practice is, in small buildings, to build non-load-bearing block partitions and load-bearing block partitions supporting one floor or a roof off the oversite concrete which serves as a foundation for them. If there is a hardcore bed below the oversite concrete it is generally considered wise to spread a thicker bed of concrete under the partition instead of hardcore. This arrangement is illustrated in Fig. 229.

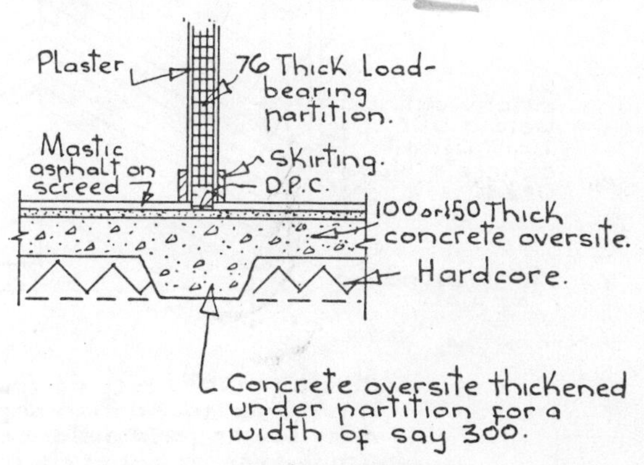


**Fig. 229**

If the partition were built on the oversite concrete over a hardcore bed its weight might cause the concrete to crack due to compaction of the hardcore.

Load-bearing partitions of brick or concrete blocks which aid in supporting two or more floors and a roof carry such loads at their base that it is economical to form a foundation for them on firm subsoil below the surface. The loads on the base of the partition and the nature of the subsoil determine at what level the foundation can most economically be formed as explained in Vol. 1, Chapter 1. The width of strip concrete foundation and its thickness are determined for partitions as they are for walls. The foundation of a partition is formed on a stratum of firm subsoil as close to the surface as possible to economise in excavation and foundation work. The base of the foundation of the partition can often be above the base of the wall foundation and it is necessary therefore to unite the two. This is done by forming a step in the foundation of the partition as illustrated in Fig. 230.

As with external walls the brickwork below ground must be laid in cement mortar up to the horizontal damp proof course.

There must be a horizontal D.P.C. in the partition and it must lap 150 over or under the D.P.C. in the external walls. Any one of the materials described in Volume 1, Chapter II may be used.

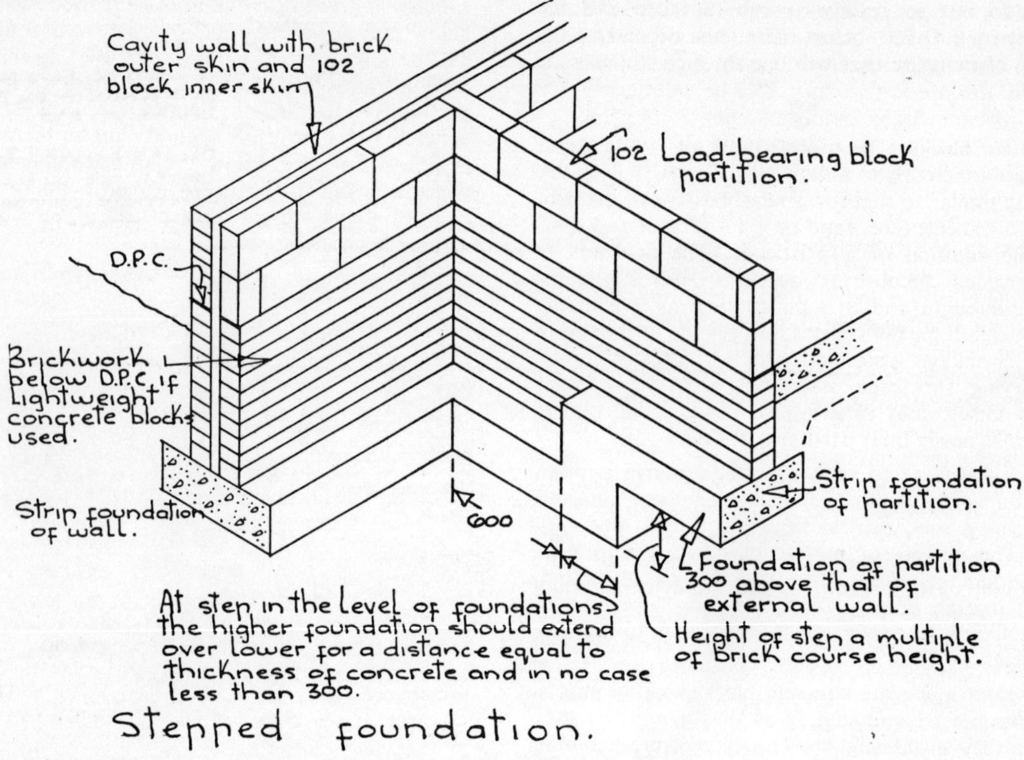
Cavity wall with brick outer skin and 102 block inner skin
102 Load-bearing block partition.
D.P.C.
Brickwork below D.P.C. if Lightweight concrete blocks used.
Strip foundation of wall.
600
Strip foundation of partition.
Foundation of partition 300 above that of external wall.
At step in the level of foundations the higher foundation should extend over lower for a distance equal to thickness of concrete and in no case less than 300.
Height of step a multiple of brick course height.
Stepped foundation.

Fig. 230

**Blocks below ground:** Dense aggregate concrete blocks and hollow clay blocks can be used in building the foundation of load-bearing partitions below D.P.C. level as they do not readily absorb moisture and are not therefore liable to damage by frost.

None of the lightweight aggregate concrete blocks can safely be used for building partitions or walls below ground because they very readily absorb moisture and are therefore liable to damage by frost.

If lightweight aggregate concrete blocks are used in a partition which requires a foundation below the level of the oversite concrete, then that part of the partition below D.P.C. level should be built with bricks in cement mortar.

**Damp proof course:** It is good practice to sandwich a waterproof membrane of bitumen or pitch in the oversite concrete under solid ground floors formed on soils liable to retain moisture—Volume 1, Chapter V. If a partition is built on oversite concrete with a waterproof membrane in it there is no necessity to form a separate D.P.C. at the base of the partition. On dry subsoils such as well drained gravel, ballast and sand a layer of oversite concrete without a waterproof membrane is generally sufficient to prevent any moisture rising through it, and lightweight block partitions are built on it without a D.P.C.

If the surface of a solid ground floor is finished with mastic asphalt it is common practice not to sandwich a waterproof membrane in the site concrete, even on subsoils that retain moisture, as the asphalt will prevent damp rising into the building. But a partition built directly off the site concrete may well absorb some moisture from the subsoil through the concrete. It would seem logical to run the asphalt right across the floor and then build the partition off it. But if this were done the surface of the asphalt would almost certainly be damaged during block laying. Instead the partition should be built on a continuous D.P.C. such as lead cored felt and the asphalt floor finish run up to the D.P.C. as shown in Fig. 229. Thermoplastic tiles and wood blocks are bedded in a thin layer of bitumen which is often considered sufficient to prevent moisture rising from site concrete. Again if there is any likelihood of moisture rising through the site concrete a D.P.C. should be laid under a partition built on it and the bitumen bed for thermoplastic tiles or wood blocks run up to it. Lightweight concrete blocks very readily absorb moisture and if a partition is built with them on oversite concrete without a waterproof membrane in it it is wise to lay a D.P.C. below the partition if there is any likelihood of moisture rising through the concrete.

Dense aggregate concrete blocks and hollow clay

blocks do not so readily absorb moisture and the necessity for a D.P.C. below them must depend on the amount of moisture that will rise through the oversite concrete.

**Mortar for blocks:** The mortar used for laying dense aggregate concrete and hollow clay blocks is generally of a mix similar to that for general brickwork, namely 1 : 1 : 6 cement, lime, sand or 1 : 4 cement and sand with the addition of a plasticiser. The first mix is recommended because it develops sufficient compressive strength and it is plastic, and as it contains only an eighth part of cement is not liable to the severe drying shrinkage associated with rich cement mixes. The drying shrinkage associated with rich cement mortar mixes does often cause cracking of plaster surfaces as newly built partitions dry out.

The 1 : 4 cement to sand mix noted above is often preferred by bricklayers as it is obviously less laborious to prepare a mix with two ingredients than one with three. The makers of mortar plasticisers claim that their product controls shrinkage as well as making cement mortars plastic. Plasticisers have not been in use a sufficient length of time for experience to bear out their claim.

Lightweight aggregate concrete blocks readily absorb moisture and expand slightly as they do so. As they dry out they shrink slightly. This type of block should be protected from excessive wetting on site as explained in Vol. 1, Chapter XI. Because of this moisture movement lightweight concrete blocks should be laid in a mortar with roughly the same porosity and drying shrinkage as they have. Rich cement mortars should be avoided. The mix of mortar recommended for this type of block is 1 : 2 : 9 of cement, lime and sand. With this mix of mortar the shrinkage of the blocks and the mortar on drying out is about the same and cracking of plastered surfaces is unlikely.

**Bonding blocks to brickwork:** Where block partitions are built up to brick walls or partitions the block work must in some way be bonded to the brickwork to improve its vertical stability. One of the more usual ways of bonding blocks to brickwork is to form pockets or recesses in the brickwork as it is built by bedding bricks on end every alternate three courses. Into the 235 high by 122·5 mm wide by 37·5 mm deep pockets alternate courses of blocks can be bonded as shown in Fig. 231. This is a perfectly satisfactory and quickly executed way of bonding the blocks to the brickwork. But some local authorities forbid the use of this method of bonding because the pockets in the brickwork reduce its strength. They might just as logically forbid window openings in walls. If it is not possible to form pockets in brickwork the blocks are bonded to it by means of cavity wall ties or strips of expanded metal built into the brickwork every third course and projecting from it for bonding to the blocks as illustrated in Fig. 232.

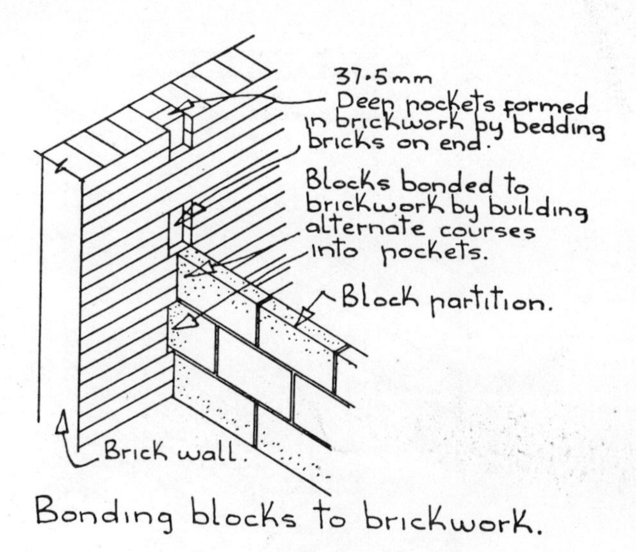


**Fig. 231**

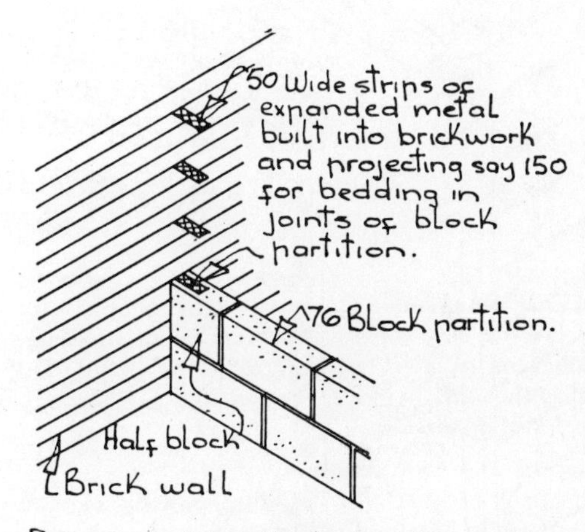


**Fig. 232**

**Bonding block walls together:** Where block partitions intersect or return at an angle they should be bonded as shown in Fig. 233.

**Door opening in block partitions:** Non-load-bearing partitions 51 and 64 thick are not particularly stable and to form an opening in them without strengthening the partition is not good practice. Storey height door frames should be built into the partition. A storey height door frame is made so that it can be fixed top and bottom between floor and ceiling as illustrated in Fig. 234.

The blocks are built up to and into the rebate in the back of the storey height frame and this helps to strengthen the partition.

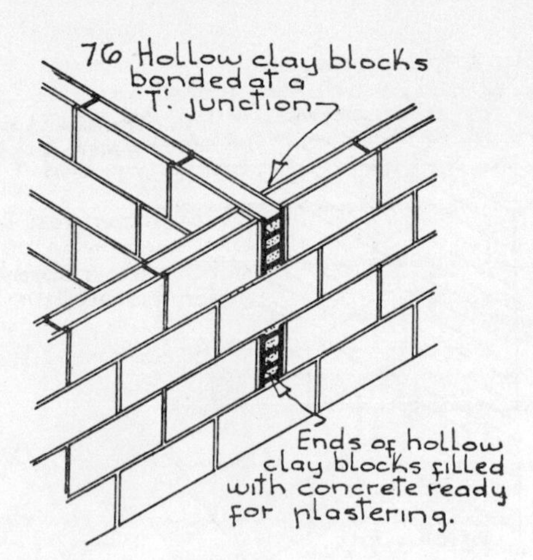


Fig. 233

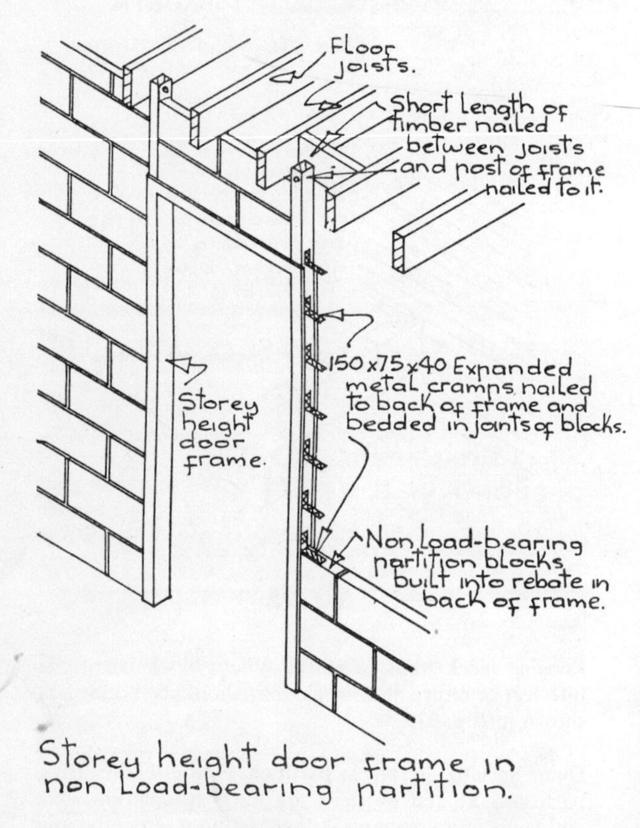


Fig. 234

It is good practice to nail strips of expanded metal laths or strips of copper sheets 150 x 75 x 25 to the back door frames and linings built into block partitions. The strips are built into horizontal joints between blocks and serve to bind frames and linings to partitions.

Openings for doors in partitions 76 or more thick can generally be formed without the need to strengthen the partition. The blocks over the door opening can be supported on a timber, concrete or block lintel. Timber lintels are of the same thickness as the blocks they support, 75 or 100 deep and built into the jambs of the partition at least 75 each end.

Concrete lintels are usually precast the same thickness as the blocks they are to support. They are made 75 deep for openings 600 wide and cast without reinforcement, and 150 deep for openings up to 900 wide with one 10 reinforcing bar. The ends of these lintels are usually built in 75 to 150.

Hollow concrete or hollow clay blocks can be used as a lintel by placing two or more of them on end, threading a reinforcing bar through a cavity and filling the cavities with concrete. When the concrete has set they serve as a lintel and are raised and built in over door openings, as illustrated in Fig. 235.

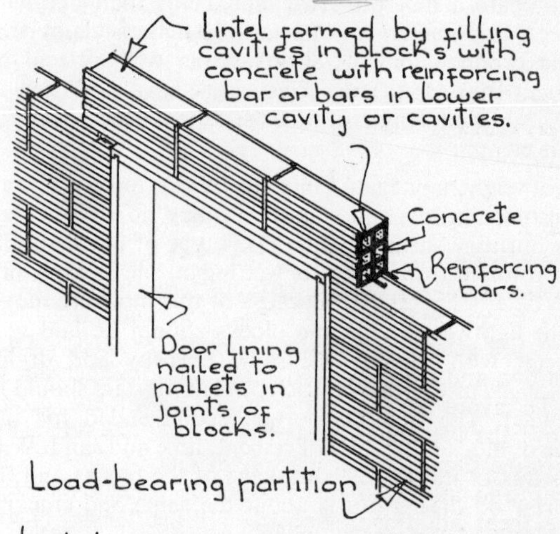


Fig. 235

**Support of block partitions on upper floors:** In planning the layout of rooms on upper floors partitions often do not lie above partitions in the floor below and the upper floor has to be made strong enough to carry the weight of the partitions built on it and superimposed loads. If non-load-bearing partitions are built on reinforced concrete upper floors and are not given support by load-bearing partitions below them the engineer designing the floor has to allow for this additional load in calculating the stress and reinforcement of the floor. Non-load-bearing partitions built on timber upper floors and not supported by a load-bearing partition in the floor below are supported by strengthening the timber floor. If the length of the partition is parallel to the span of the floor joists it is generally supported by two floor joists as illustrated in Fig. 236.

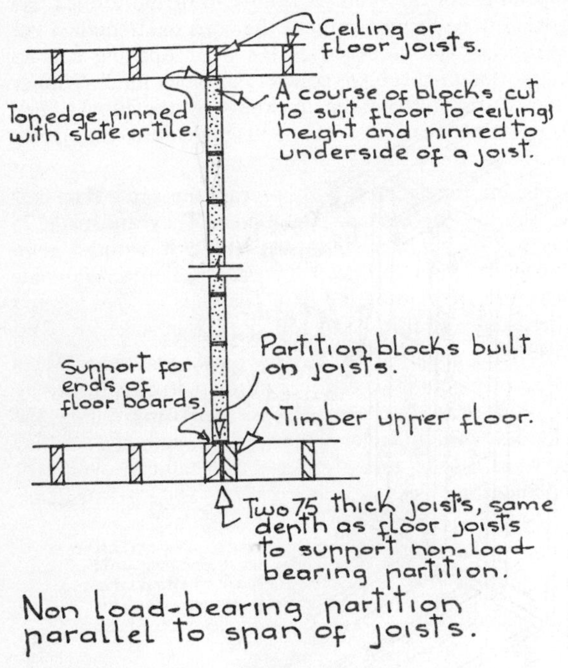


Fig. 236

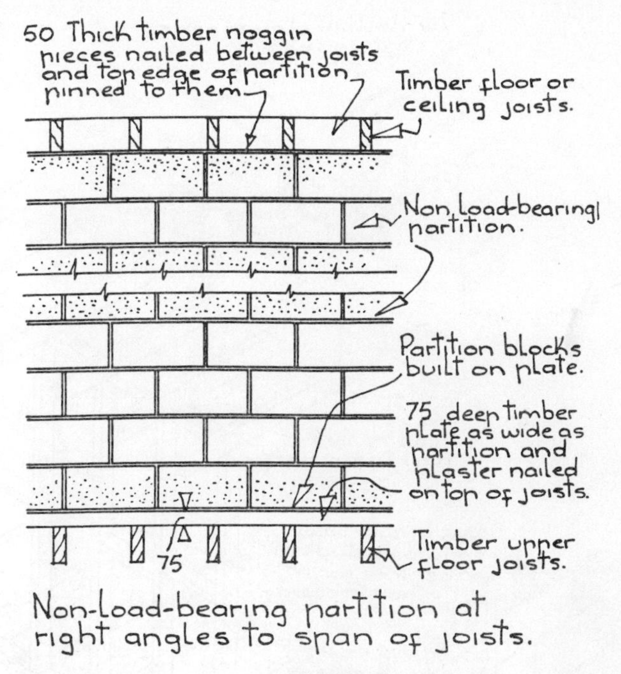


Fig. 237

When fixing the floor joists the carpenter first places the two joists in position to carry the partition and then spaces the joist each side at equal intervals. Two joists are generally necessary to carry the load of the partition and the ends of the floor boards either side of it. To avoid an unsightly projection in the ceiling below, the two joists below the partition should be of the same depth as the rest of the floor joists around them, and are often made 75 thick instead of 50 to give them additional strength. If the length of the partition is at right angles to the span of the joists it is supported either by spacing the joists more closely than usual (say 300 centres) or by increasing the depth of the joists or both. The partition is then built off a timber plate nailed across the top of the floor joists as illustrated in Fig. 237.

**Pinning top edges of partitions:** The top of non-load-bearing partitions should be pinned to the underside of floors or ceilings over them. If they are built up to the underside of concrete floors or roofs the top of the partition is pinned with slate or tile slips in mortar wedged firmly into the space between the top course of blocks and the soffit above. Similarly if the partition lies under a timber joist parallel to its length, wedges of slate or tile in mortar are used as illustrated in Fig. 236. If the length of the partition is at right angles to the timber joists above, its top edge may be cut around and pinned to the joists or better still timber noggins are fixed between the joists to which the partition is wedged as illustrated in Fig. 237.

**Timber stud partitions:** A stud partition consists of 100 x 50 or 75 x 50 sawn softwood timbers fixed vertically between a timber plate and head fixed at floor and ceiling level respectively as in Fig. 238.

The plate and the head of the partition are of the same section as the studs which are fixed between them, with the 50 wide face of studs showing on the faces of the partition. The studs are nailed to the head and plate and the head and plate are nailed to the timber soffit and floor respectively. The studs are spaced at up to 450 intervals if they are to be faced with lath and plaster and 400 if they are to faced with plaster-board.

Studs of 75 x 50 are used for partitions up to 2·4 in height and 100 x 50 for partitions up to 3·0 in height.

The short lengths of timber, called noggin pieces, nailed between the studs are an essential part of the partition as they prevent the studs from bowing out of vertical and from twisting due to shrinkage. Without the noggin pieces slight twisting or bowing of the studs would cause plaster on the partition to crack. The noggin pieces may be fixed in line, staggered or herringbone between the studs as illustrated in Fig. 239. Accurately cut noggin pieces fixed herringbone are more effective than either the in-line or staggered method. Whichever method of fixing the noggins is used their ends should be firmly wedged between walls as illustrated in Fig. 238.

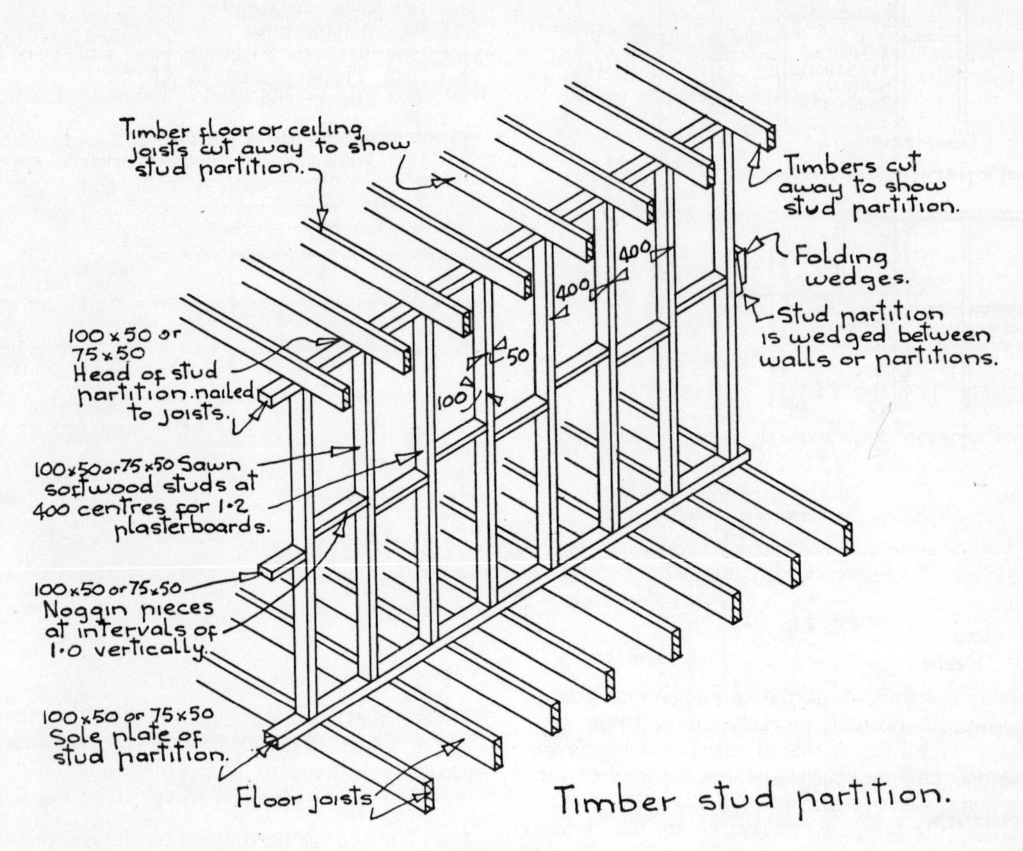


Fig. 238

**Openings in stud partitions:** The simplest method of forming an opening in a stud partition is to fix studs each side of the opening and nail a short length over as the head of the opening as illustrated in Fig. 240. But this is not an entirely satisfactory method as the studs are not rigidly square-framed around the opening and may move, so causing the door to jam in its linings or frame. The better method of forming the opening is to frame slightly larger section timbers around the opening as illustrated in Fig. 240.

It will be seen that 100 x 75 or 75 x 75 timbers are used and jointed rigidly with a wedged mortice and tenon-joint which secures them at right-angles, one to the other.

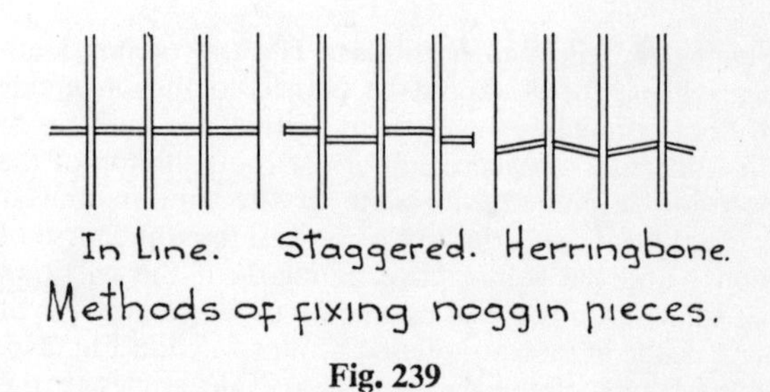


Fig. 239

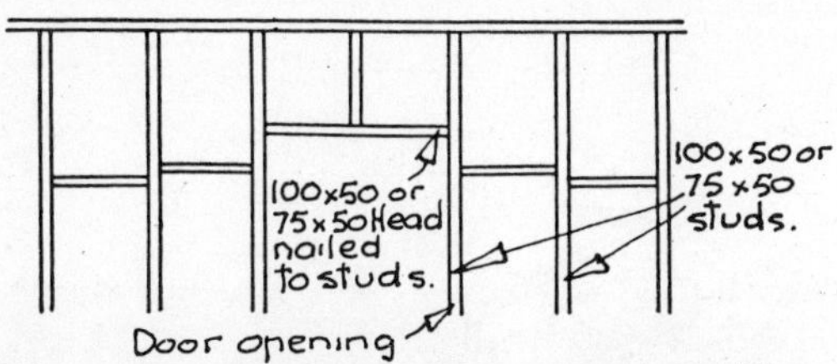


Door opening trimmed in stud partition.

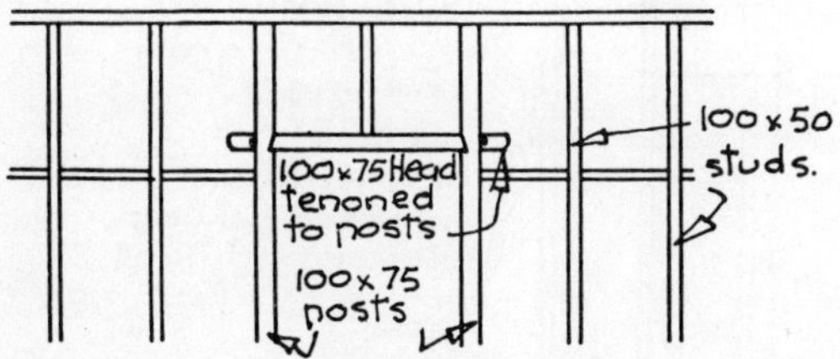


Door opening framed in stud partition.

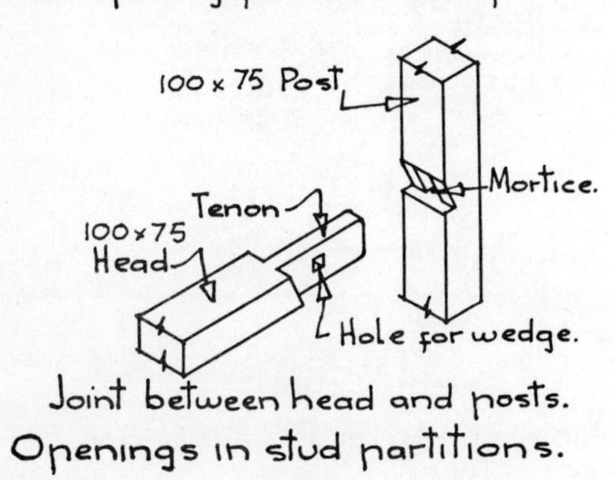


Joint between head and posts.

Openings in stud partitions.

**Fig. 240**

### Reference Books and Publications

**British Standards:**
No. 12. Portland cement.
No. 890. Building limes.
No. 1200. Sands for brickwork, block walling and masonry.
No. 2028. Precast concrete blocks.
No. 3921. Bricks and blocks of fired brickearth, clay or shale.
**British Standard Code of Practice:**
CP. 122. Walls and partitions of blocks and slabs.

CHAPTER SEVEN

# STAIRS

## SfB (24) Stairs: General

A stair may be constructed with steps rising without break from floor to floor, or with steps rising to a landing between floors, with a further series of steps rising from the landing to the floor above.

**Flight:** The word flight describes an uninterrupted series of steps between floors, or between floor and landing, or between landing and landing. A flight should not have more than 14 or 15 steps in it, otherwise it might be dangerous for the elderly or young. The widths of flights and landings should be not less than 850 measured overall. The usual width of flights and landings for stairs in houses is 900.

**Landing:** A landing is a level platform constructed between floors either where flights of steps change direction, or to make a break in what would otherwise be an over long flight of steps.

**Tread and Riser:** The horizontal face of a step is described as the tread. The vertical, or near vertical, face of a step is described as the riser. Treads commonly project beyond the faces of risers as a nosing to provide as wide a surface of tread as practicable. Fig. 241 illustrates the use of the descriptions tread, riser and nosing.

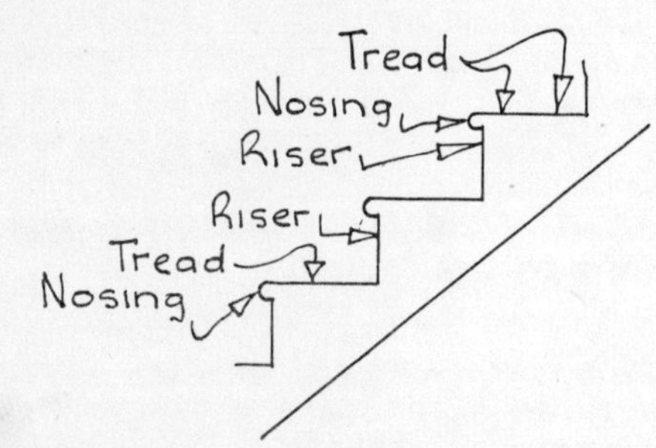


**Fig. 241**

**Rise and Going:** The word "rise" describes the distance, measured vertically, from the surface of one tread to the surface of the next.
The word "going" describes the distance, measured horizontally, from the face of one riser to the face of the next riser. Fig. 242 illustrates the use of these terms diagrammatically, and gives the accepted minimum headroom over flights.

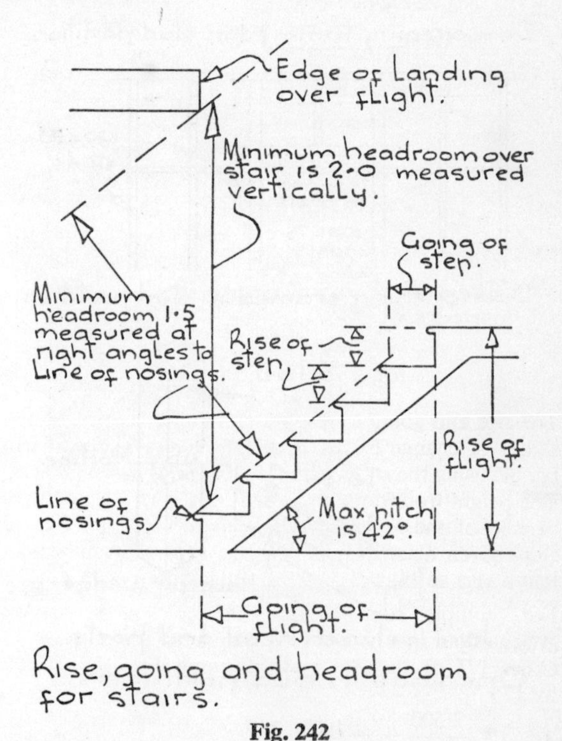


**Fig. 242**

The dimensions of the rise and going of its steps determine whether a stair is steep or shallow. For example if the rise of a stair is 75 and the going 300 it will be shallow and if the rise is 225 and the going 100 it will be very steep as illustrated in Fig. 243. The shallow stair illustrated would be tedious and the steep one almost impossible to climb. So by tradition various dimensions of rise and going have been accepted.

For houses the accepted dimensions are:
Going: not less that 210-225
Rise: between 175 and 200
For stairs used in common by occupiers of flats:
Going: 225 to 250
Rise: not more than 190
For stairs to public buildings:
Going: 280 Rise: 150

It will be seen that the accepted dimensions of rise and going for houses produce a steeper stair than those accepted for public buildings. The steeper stair is accepted for houses because a shallow stair would occupy more area on plan and thus reduce living area.

(The shallower a stair the more risers required for given height and therefore the more treads required.) The shallow stair recommended for public buildings is designed to minimise danger to the public escaping via the stairs during emergency.

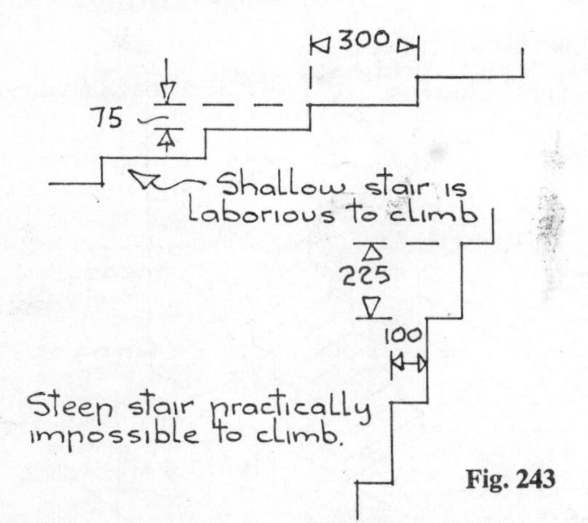


Fig. 243

The rise and going of every step in a stair must be the same. A change of rise or going is very likely to trip people using the stair, with possible fatal consequences. The height from floor to floor is not always a multiple of one of the accepted whole figures for the rise of steps given above. It is necessary therefore to calculate a rise so that each step is the same height.

For example, suppose first floor level is 2·5 above ground floor in a house. This height is not a multiple of either 175 or 190. To find a suitable rise divide 2·5 by 175

$$\frac{2{\cdot}500}{175} = 14\tfrac{50}{175}$$

Obviously if the rise of every step is to be the same, there cannot be $14\frac{50}{175}$ of them. If there are 15 risers each will have a $\frac{2{\cdot}500}{15} = 166\frac{2}{3}$ rise if there are 14 each will have $\frac{2{\cdot}500}{14} = 178\frac{4}{7}$ rise. Using 225 going and either 14 or 15 risers in this example will give quite a comfortable stair.

**Types of stair:** The three basic ways in which stairs are planned are as:

(1) A straight flight stair.
(2) A quarter turn stair.
(3) A half turn stair.

A straight flight stair rises from floor to floor in one direction, with or without a landing, hence the term "straight" flight.

A quarter turn stair rises to a landing between floors, turns through 90 degrees, then rises to the floor above, hence 'quarter turn'.

A half turn stair rises to a landing between floors, turns through 180 degrees, then rises, parallel to lower flight, to floor above, hence 'half turn'.

Fig. 244 illustrates the three basic arrangements of stairs diagrammatically.

The type of stair used depends mainly on the layout of the rooms in a building. A straight flight stair was commonly used in country cottages and is sometimes known as a cottage stair.

A half turn stair is often described as a "dog leg stair" because it looks somewhat like a dog's hind leg in section. (See Fig. 244).

Stairs are sometimes described as "open well stairs." The description refers to a space or well between flights. A half turn stair can be arranged with no space between the flights or with a space or well between them and this latter arrangement is sometimes described as an open well stair. A quarter turn stair can also be arranged with a space or well between the flights when it is also an open well stair. As the term "open well" does not describe the arrangement of the flights of steps in a stair, it should only be used in conjunction with the more precise descriptions straight flight, quarter or half turn stair (e.g. half turn stair with open well).

**Staircase**

Stairs in houses are usually constructed from timber boards put together in the same way as a box or case, hence the term staircase.

Each flight of a staircase is made up (cased) in a joiners shop as a complete flight of stairs. Landings are constructed on site and the flight or flights are fixed in position between landings. The members of a staircase flight are strings (or stringers), treads and risers. The treads and risers are joined to form the steps of the flight and are housed in, or fixed to strings whose purpose is to support them. Because the members of a flight are put together like a box, thin boards can be used and yet be strong enough to carry the loads normal to stairs.

The members of a flight are usually cut from the timbers of the following sizes:

Treads 32 or 38 Risers 25 or 19 Strings 38 or 50

Fig. 245 is a view of a flight of a staircase, with some of the treads and risers taken away to show the housings in the string into which they fit.

**Joining risers to treads:** One method of joining risers to treads is to cut tongues on the edges of the risers and fit them into grooves cut in the treads as illustrated in Fig. 246 on page 120.

Another method is to butt the top of the riser under the tread with the joint between the two, which would otherwise be visible, masked by a moulded bead as illustrated in Fig. 246.

The tread of a stair tends to bend under the weight of people using it. When a tread bends, the tongue on

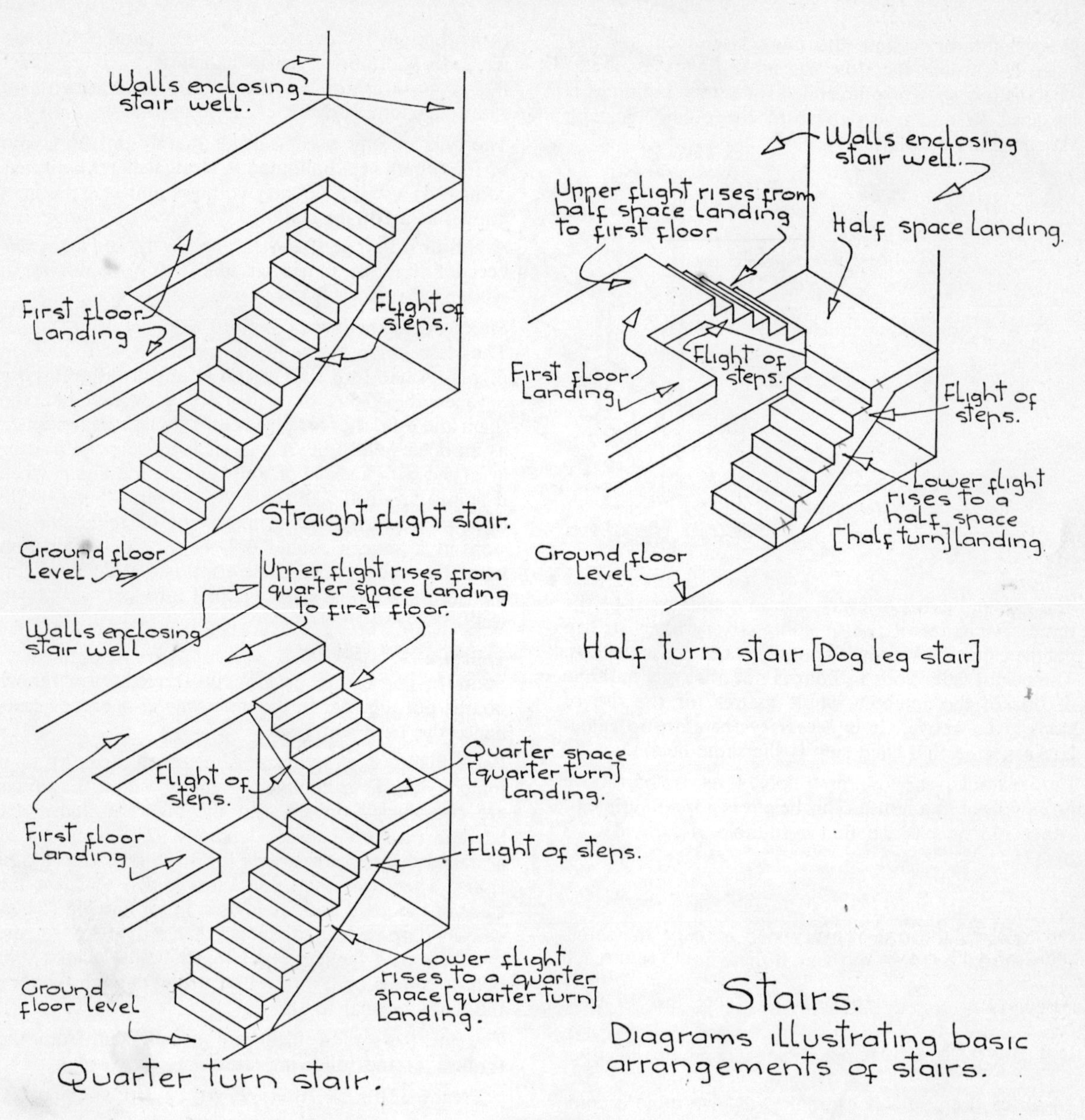


Fig. 244

the bottom of the riser comes out of the groove in the tread, and the staircase "creaks". To prevent this objectionable creaking it is common practice today to secure the treads to risers with screws as illustrated in Fig. 246.

**Nosing on treads.** The nosing on treads usually projects 32, or the thickness of the tread, from the face of the riser below. A greater projection than this would increase the likelihood of the nosing splitting away from the tread and a smaller projection would reduce the width of tread.

The nosing is rounded for appearance sake. Fig. 246. illustrates the more usual finishes to nosings.

**Strings.** Strings (stringers) are cut from boards 38 or 50 thick and of sufficient width to contain and support the treads and risers of a flight of steps.

Staircases are usually enclosed in a stair well. The stair well is formed by an external wall or walls and brick or block partitions, to which the flights and landings are fixed. The string of a flight of steps which is fixed against a wall or partition is termed the wall string, and the other string of the same flight, the

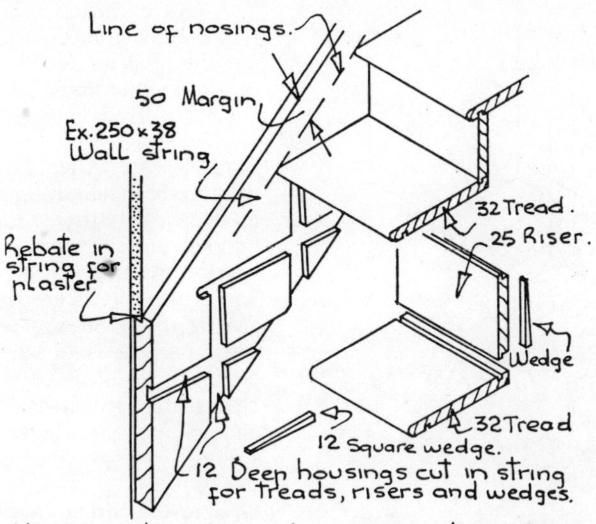
Line of nosings.
50 Margin
Ex. 250 x 38
Wall string
Rebate in string for plaster
32 Tread.
25 Riser.
Wedge
32 Tread
12 Square wedge.
12 Deep housings cut in string for treads, risers and wedges.

Housing treads and risers in close string.

**Fig. 247**

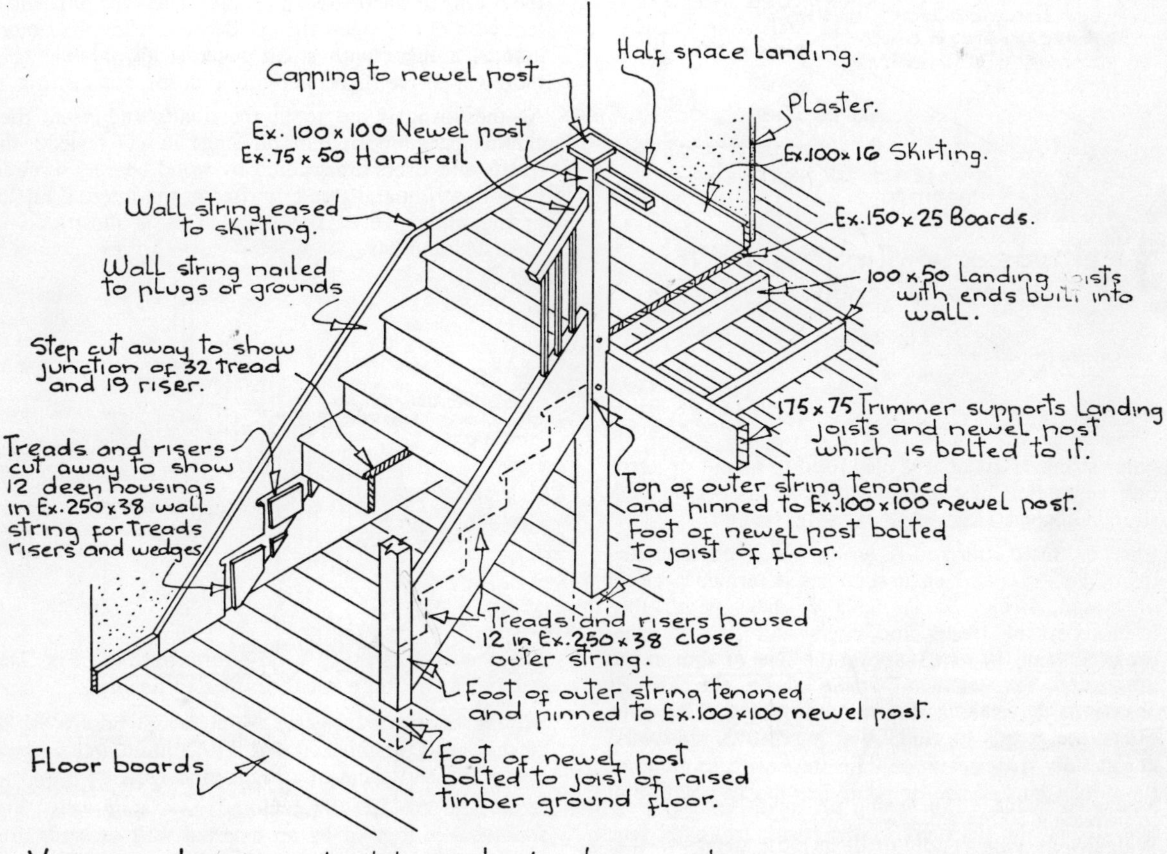
Half space Landing.
Capping to newel post
Plaster.
Ex. 100 x 100 Newel post
Ex. 75 x 50 Handrail
Ex. 100 x 16 Skirting.
Wall string eased to skirting.
Ex. 150 x 25 Boards.
Wall string nailed to plugs or grounds
100 x 50 Landing joists with ends built into wall.
Step cut away to show junction of 32 tread and 19 riser.
175 x 75 Trimmer supports Landing joists and newel post which is bolted to it.
Treads and risers cut away to show 12 deep housings in Ex. 250 x 38 wall string for treads risers and wedges
Top of outer string tenoned and pinned to Ex. 100 x 100 newel post.
Foot of newel post bolted to joist of floor.
Treads and risers housed 12 in Ex. 250 x 38 close outer string.
Foot of outer string tenoned and pinned to Ex. 100 x 100 newel post.
Floor boards
Foot of newel post bolted to joist of raised timber ground floor.

View of lower flight of half turn staircase.

**Fig. 245**

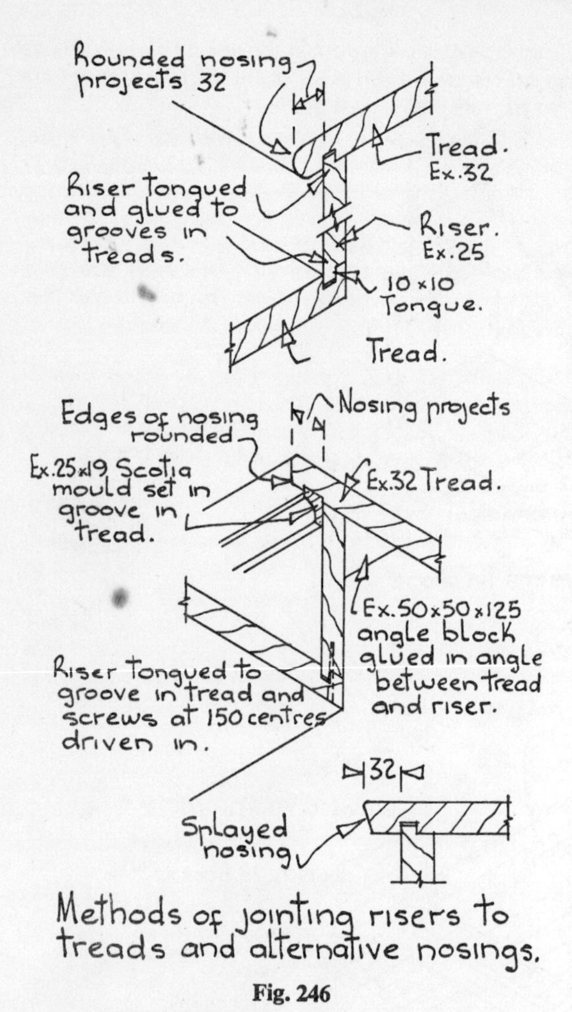


**Fig. 246**

outer string. (Unless it is also fixed to a wall or partition when it is also a wall string. This occurs with straight flight stairs.)

**Close or closed strings:** A string which encloses the treads and risers which it supports is termed a close, or closed, string. It is made of such width that it encloses the treads and risers and its top edge projects some 50 or 63 above the line of the nosing of treads. The width of string above the line of nosings is described as the margin. Fig. 247 on page 119 shows a close string. A string 250 or 280 wide is generally sufficient to contain steps with any one of the dimensions of rise or going previously noted, and to provide a 50 margin.

Wall strings are generally made as close strings so that wall plaster can be finished down on to them.

Outer strings can be made as closed strings or as open (cut) strings.

The ends of treads and risers are glued and wedged into shallow grooves cut in closed strings. The grooves are cut 12 deep into the strings and tapering slightly in width to accommodate treads, risers and the wedges which are driven in below them as illustrated in Fig. 247.

**Angle blocks.** After the treads and risers have been put together and glued and wedged into their housings in the closed strings, angle blocks are glued in the internal angles between the underside of treads and risers, and underside of treads and risers and strings. Angle glue blocks are triangular sections of softwood cut from say 50 square timber and each 120 long. Their purpose is to strengthen the right-angled joints between treads, risers and strings. Three or four blocks are used at each junction of tread and riser and one between the end of each tread or riser and string.

**Open or Cut String.** A closed outer string looks somewhat lumpy and does not show the profile of the treads and risers it encloses. The appearance of a staircase is considerably improved if the outer string is cut to the profile of the treads and risers. This type of string is termed a cut or open string. Because it involves more labour, a flight with a cut outer string is somewhat more expensive than one with a closed outer string.

As the string is cut to fit the treads and risers they cannot be supported in housings in it. Instead the treads and risers are secured by wood bearers screwed to both string and tread or riser in the internal angles in the underside of the flight. This is illustrated in Figs. 248 and 249.

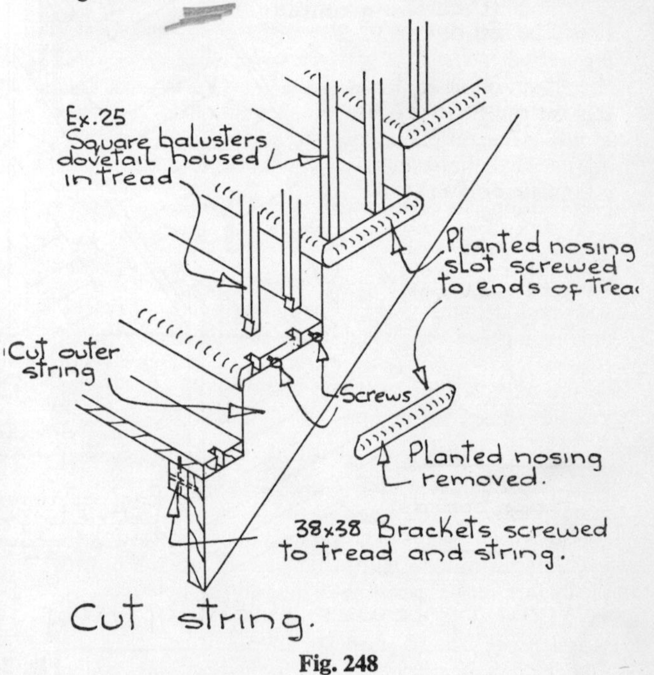


**Fig. 248**

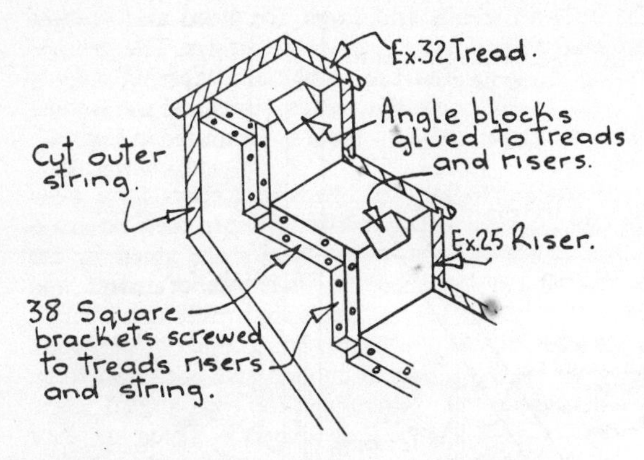

Ex.32 Tread.
Angle blocks glued to treads and risers.
Cut outer string.
Ex.25 Riser.
38 Square brackets screwed to treads risers and string.
View of underside of flight to show method of fixing treads and risers to cut outer string.

**Fig. 249**

It is not possible to cut a neat nosing on the end grain of treads to overhang the cut string, so planted nosings are fitted as illustrated in Fig. 248. The planted nosings are often secured to the ends of treads by slot screwing. This is a form of secret fixing used to avoid having the heads of nails or screws exposed. Countersunk headed wood screws are driven into the ends of treads so that their heads protrude some 12. The heads of these screws fit into holes cut in the nosing. The nosing is then knocked into position so that the heads of the screws bite into slots cut next to the holes in the nosing.

It will be seen from Fig. 248 that the planted nosings are mitred to the nosing of the tread. The end of the riser is cut off at 45 degrees to the face of the riser and this cut end fits a matching 45 degrees cut on the edge of the string. Because the string is thicker than the riser it partly butts and is partly mitred to it. This joint is sometimes termed mitre and butt.

## Landings

**Half space (turn) landing:** Is constructed with a sawn softwood trimmer which supports sawn softwood landing joists or bearers and floor boards. The trimmer is usually 175 x 75 and built into the enclosing walls or partitions of the stair well, and supports 100 x 50 landing joists as illustrated in Figs. 245 and 250 (page 122).

As well as giving support to the joists of the half turn landing the trimmer also supports a newel or newel posts. Newel posts serve to support handrails and provide a means of fixing the ends of outer strings. Fig. 250 illustrates a half turn stair with open well with the flights, newel posts and landing in position.

**Newel posts:** The newel posts are cut from 100 x 100 timbers and are notched and bolted to the landing trimmer. Mortices are cut in the newel to which tenons cut on the end of the outer string fit. These tenons are pinned with oak dowel pins.

Fig. 250 illustrates the trimmer, newel and outer string before they are assembled. Because the landing boards are usually 22 thick, a planted nosing is fixed to the edge of the landing to match the thickness of the nosings of the treads. If a half turn stair does not have a well between flights (dog-leg stair), the outer strings of both lower and upper flights are tenoned to mortices in a central newel post, one above the other.

**Drop newel:** For appearance sake the lower end of the newel post is usually finished about 100 below the soffit of flights and is moulded. As it projects below the stair it is termed a newel drop (Fig. 250).

## Balustrade

**Open balustrade:** The traditional balustrade consists of newel posts, handrail and timber balusters as illustrated in Fig. 250. The newel posts at half turn landings and at landing at floor level are housed and bolted to trimmers as previously described. These newels are fixed in position so that the face of the risers

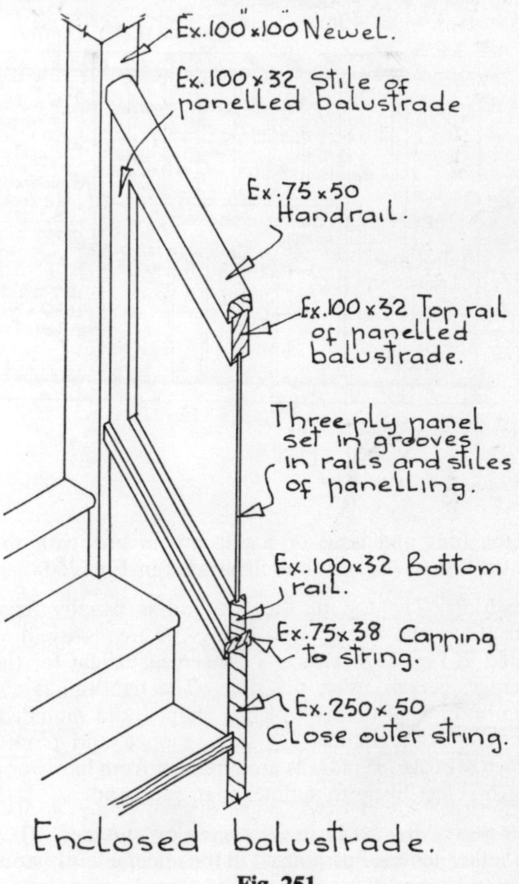


**Fig. 251**

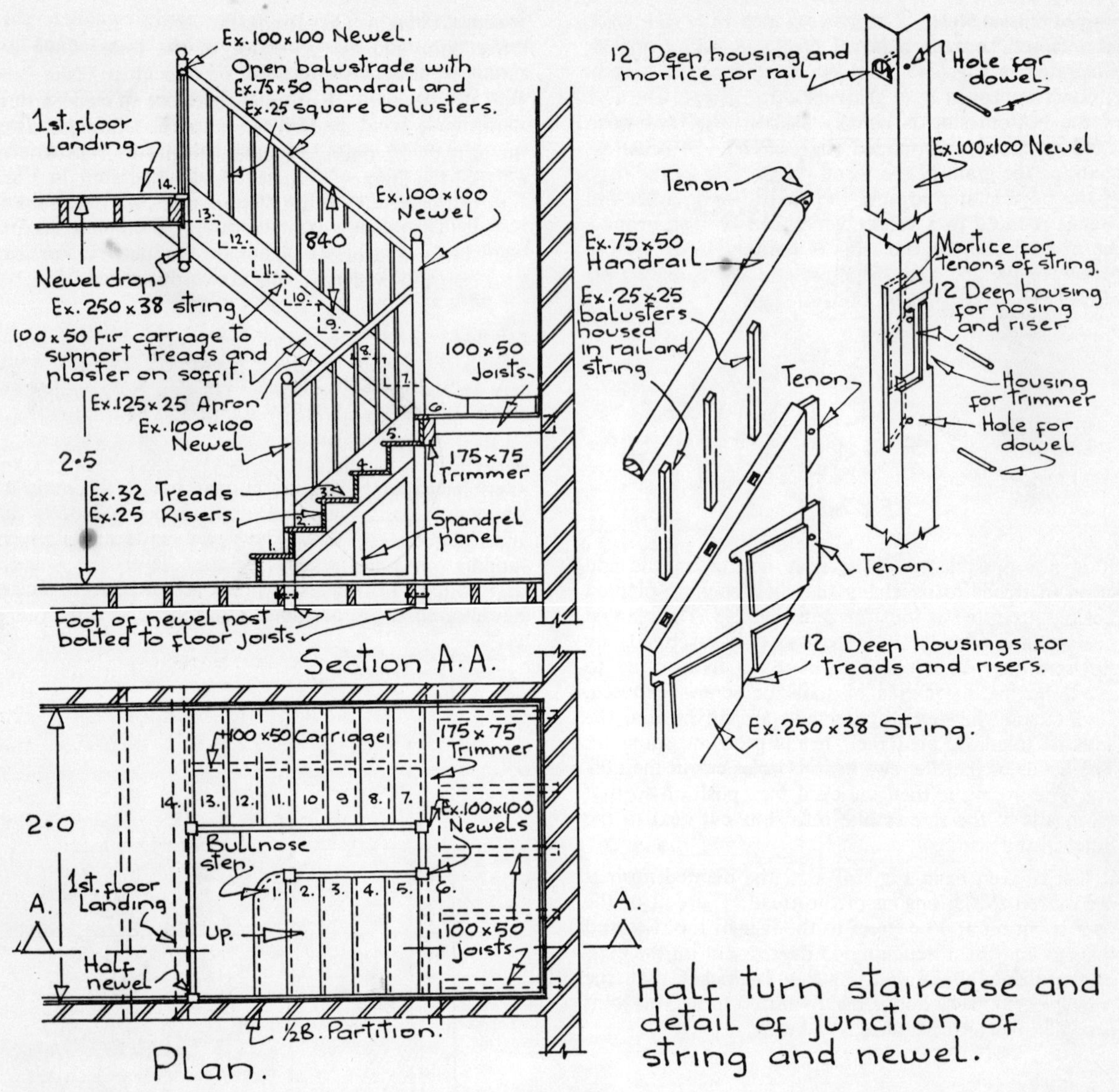


**Fig. 250**

at the foot and head of flights are in line with the centre line of the newel as illustrated in Fig. 250.

**Handrail:** The top of the handrail is usually fixed 850 above the line of the nosings of treads as illustrated in Fig. 250. This is a convenient height for the average person using the stair. The handrail is cut from a 75 x 50 timber which is shaped and moulded. The ends of handrails are stub tenoned and pinned to newel posts. Handrails are often cut from hardwood which is less likely to splinter than softwood.

**Balusters:** May be 25 or 19 square or moulded. They are either tenoned or housed in the underside of handrails and tenoned into the top of closed strings or set into housings in the treads of flights with cut strings. Fig. 250 illustrates the fixing of handrails and balusters.

**Closed (enclosed) Balustrade:** It is difficult to clean and dust between the balusters of an open balustrade and it is fashionable today to construct enclosed balustrades. Instead of balusters, the space between the string, handrail and newel posts is filled with timber framing supporting panels or with sawn softwood studs covered both sides with sheets of plywood or hardboard. Fig. 251 illustrates the construction of a closed balustrade.

The purpose of a cut outer string is to show the profile of the steps of a flight. There is no purpose therefore in using a cut string with a closed balustrade.

**Shaped bottom steps:** The bottom step of a stair may be enclosed between a newel post and wall string as illustrated in Fig. 245. But the appearance of a stair is much improved by a shaped bottom step. The end of the bottom step is usually shaped as a bull-nose (quarter circle) or rounded (half circle). In order to continue the grain of the wood on the face of the riser of the bottom step around the shaped end, either the riser is reduced to a veneer which can be bent around the shaped end, or the riser is entirely faced with a veneer. Figs. 252 and 253 illustrates the construction of bull-nose and rounded bottom steps.

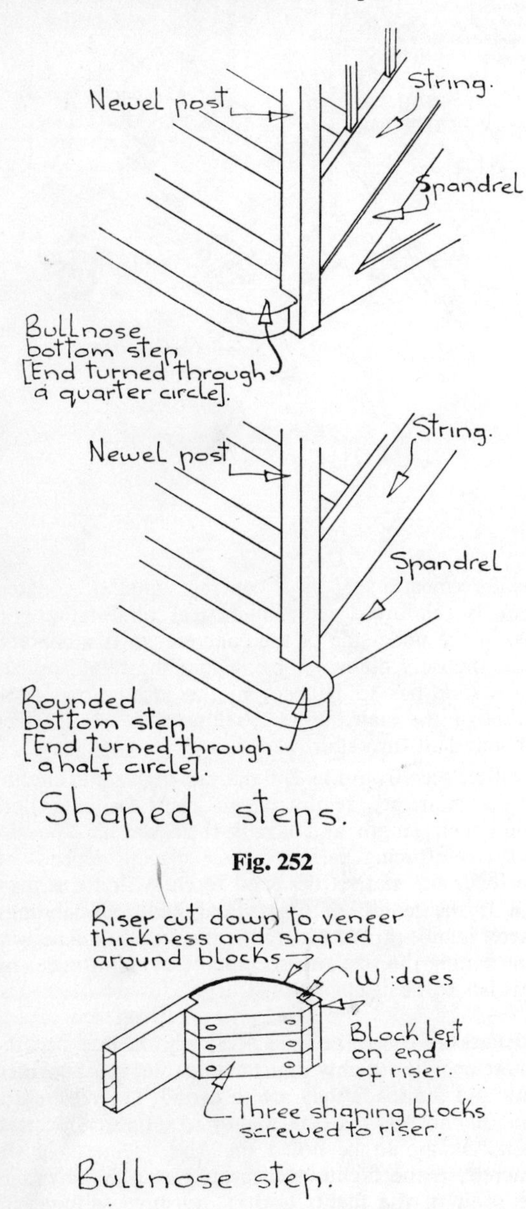


Fig. 252

Fig. 253

**Spandrel filling:** The triangular space between the lower floor and the underside of the lower flight of a stair would collect dust and be difficult to clean if it were not enclosed. It is usual therefore to enclose this space with what is termed spandrel framing. The spandrel below stairs with open balustrades is generally constructed with timber panels as illustrated in Fig. 250. The spandrel below stairs with solid balustrades may be formed as a continuation of the facing to the balustrade. To provide a fixing for the spandrel framing the newel post of the landing is usually carried down to the floor as shown in Fig. 250.

**Carriage:** After a staircase has been erected a sawn softwood carriage is fixed below the flights. This carriage serves two purposes: (a) as a background to which lath or plasterboard can be fixed for plastering on the sloping soffit of the flights and (b) as a means of strengthening the flights of steps. One 100 x 50 or 75 x 50 sawn softwood carriage is fixed centrally below each flight of steps and secured to landing trimmers or joists as illustrated in Figs. 250 and 254. To give direct support under the centre of the width of treads and risers, rough boards are cut and nailed to the side of the carriage as illustrated in Fig. 254.

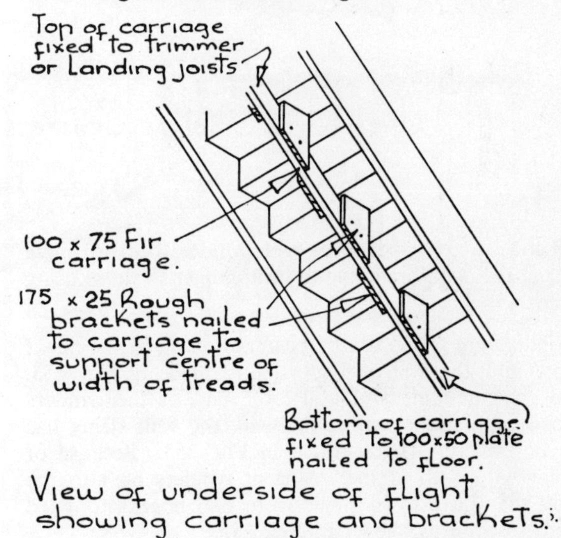

Top of carriage fixed to trimmer or Landing joists
100 x 75 Fir carriage.
175 x 25 Rough brackets nailed to carriage to support centre of width of treads.
Bottom of carriage fixed to 100x50 plate nailed to floor.
View of underside of flight showing carriage and brackets.

Fig. 254

**Quarter space (turn) landing**

A quarter space landing is supported by a newel post carried down to the floor below, as illustrated in Fig. 255. The construction of the flight, balustrade and bearers is similar to that of a half turn stair.

**Winders**

In small houses quarter or half turn stairs are sometimes constructed with winders instead of quarter or half space landings. Winders are triangular shaped steps constructed at the turn from one flight to the next as illustrated in Fig. 255. It is considered bad

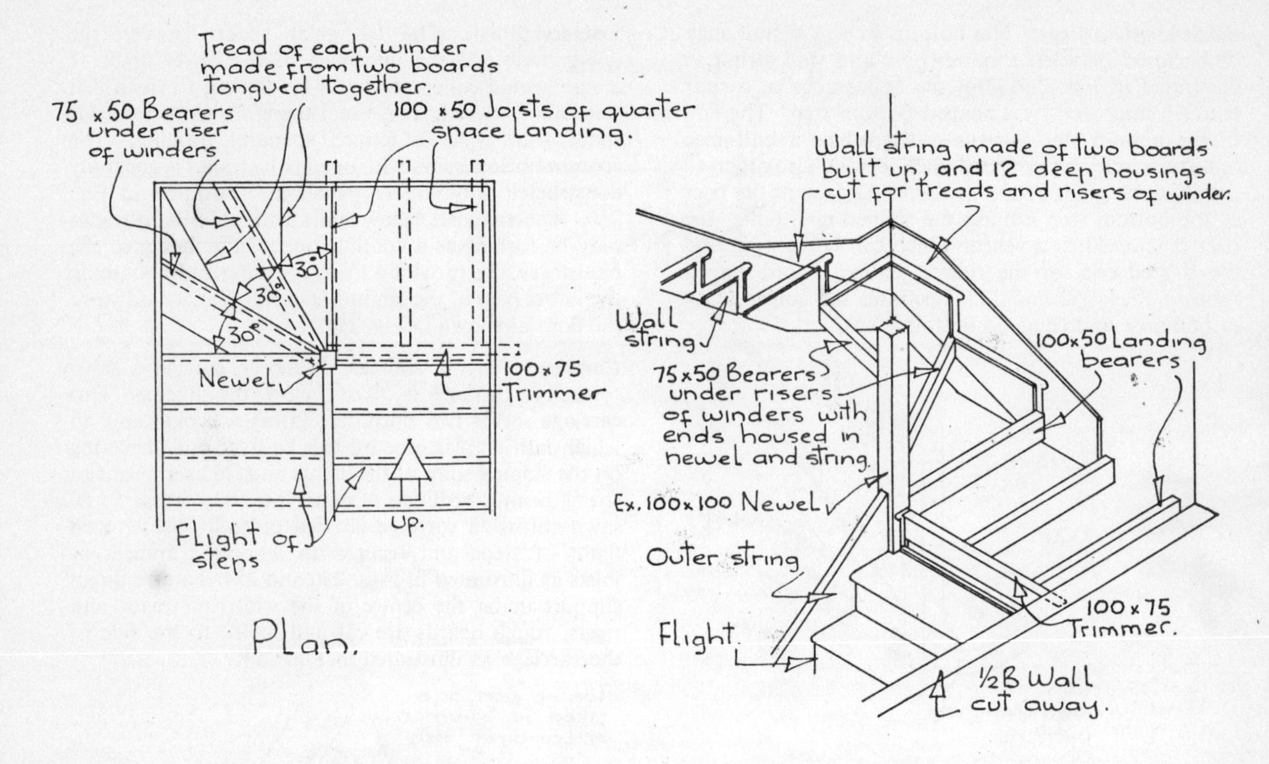


**Fig. 255**

practice to construct stairs with winders because these triangular steps are liable to be a danger to those using them.

Winders are supported by bearers housed in the newel post and the wall string as illustrated in Fig. 255. To accommodate and house the end of the winders treads and risers against the wall, the wall string has to be built up as illustrated in Fig. 255. Because of the great width of the treads of winders next to the wall they have to be made from two boards tongued and grooved as shown in Fig. 255.

### Reinforced concrete stair

A timber stair or staircase has low resistance to damage by fire. In the majority of buildings, other than two storey houses, building regulations require that stairs be capable of resisting damage by fire for one hour. Since a timber stair does not satisfy this requirement, reinforced concrete stairs are used.

The rise and going of a concrete stair are determined in the same way as for timber staircases and any arrangement of straight flight, quarter or half-turn stair may be constructed in concrete.

**Reinforcement:** Just as a concrete lintel or concrete floor is reinforced with mild steel reinforcing bars cast in the underside of the concrete, so is a concrete stair, the only difference being that the reinforcement for a stair has to be bent to the inclination of the flights of the stair. Fig. 256 illustrates a reinforced concrete half-turn stair.

It will be seen from Fig. 256 that the flights and landing of the stair are reinforced to resist bending both along their length and across their width. The size of the reinforcing bars and their spacing depends on the loads the stair is designed to carry and the span, that is the length of flights and width of landings. There is no satisfactory rule of thumb method of determining the size and spacing of the reinforcement. This has to be calculated.

**Thickness of concrete:** The necessary thickness of the concrete of the flights and landings depends on their span and the loads they are to carry. The concrete is generally not less than 100 and up to 150 thick for most stairs. It should be noted that the thickness of the concrete in the flights is measured at right angles to the soffit of the flights and is measured without including the steps.

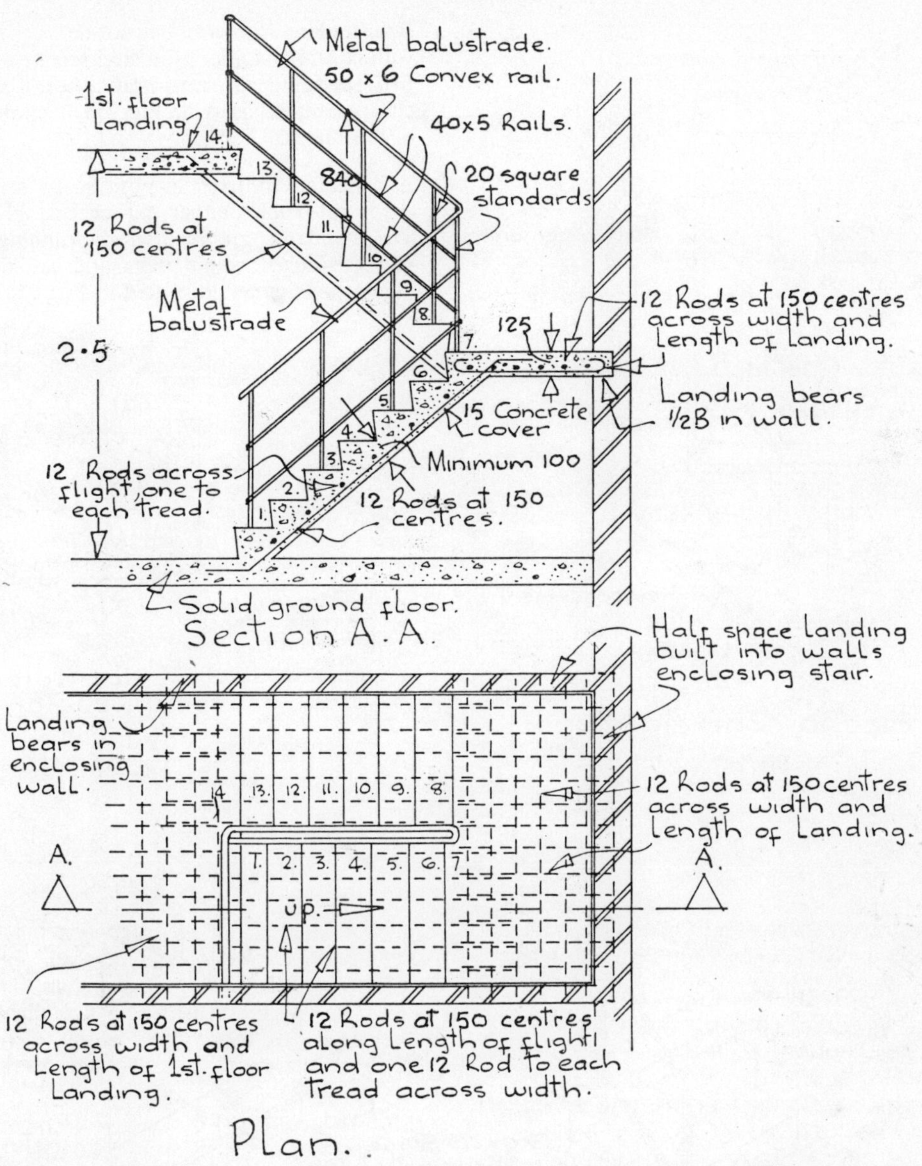


Fig. 256

**Cover to reinforcement:** As with any reinforced concrete part of a building there must be sufficient cover of concrete to the reinforcement to protect it from damage by fire and rust. This cover is at least 15.

**Bearing of stair:** Obviously a part or parts of a concrete stair must be built into walls or on offsets in walls or partitions so that the weight of the stair is carried on some load bearing part of the structure. One of the simplest ways of effecting this is to build the ends of half or quarter space landings, and landings at floor level, into walls and to design the flights to span from landing to landing without side support as illustrated in Fig. 256. The advantage of this type of construction is that it avoids the expensive, laborious cutting of bricks or blocks necessary if one side of a flight is built in.

The surface of treads and risers of concrete stairs can be finished with one of the jointless composition floor finishes, floor tiles and fittings, natural stone slabs, wood treads, or sheet rubber and metal nosings. Some typical finishes are illustrated in Fig. 257.

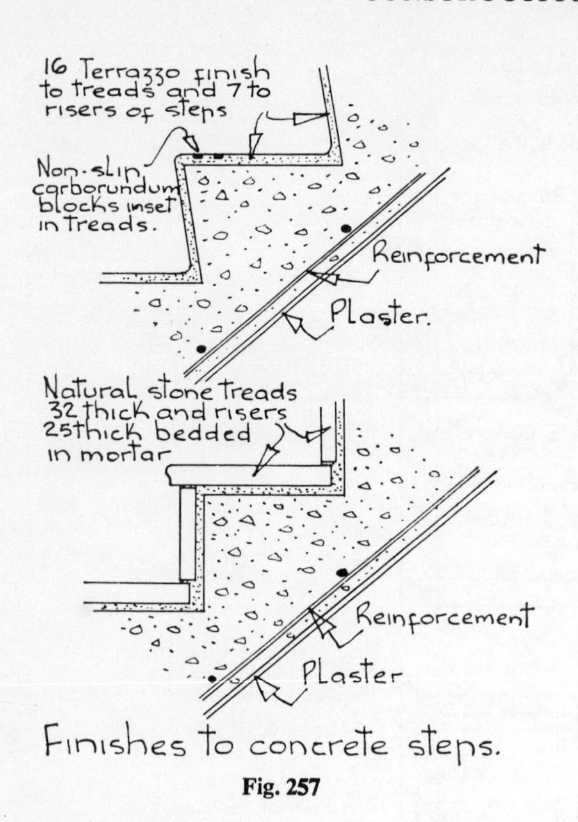


Fig. 257

**Balustrade:** A metal balustrade is commonly used with concrete stairs. Mild steel sections cut and joined to form standards, rails and core rail are one of the cheapest materials used. Fig. 256 illustrates a balustrade of this material.

A balustrade is fixed in position by setting the standards into holes either cut or cast in the concrete. The standards are secured by running molten lead into the holes around them and caulking (ramming) it into position as illustrated in Fig. 258.

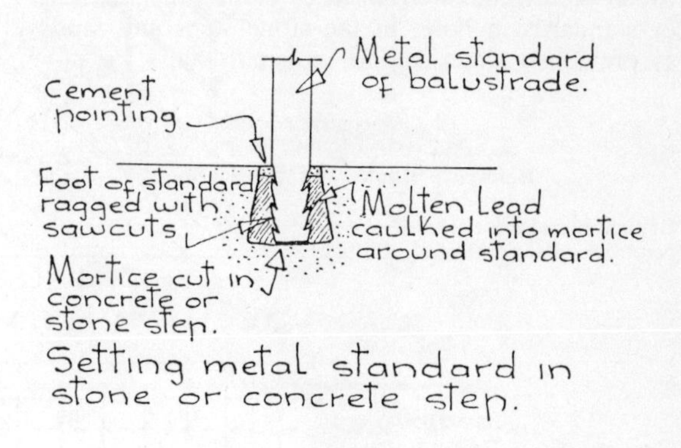


Fig. 258

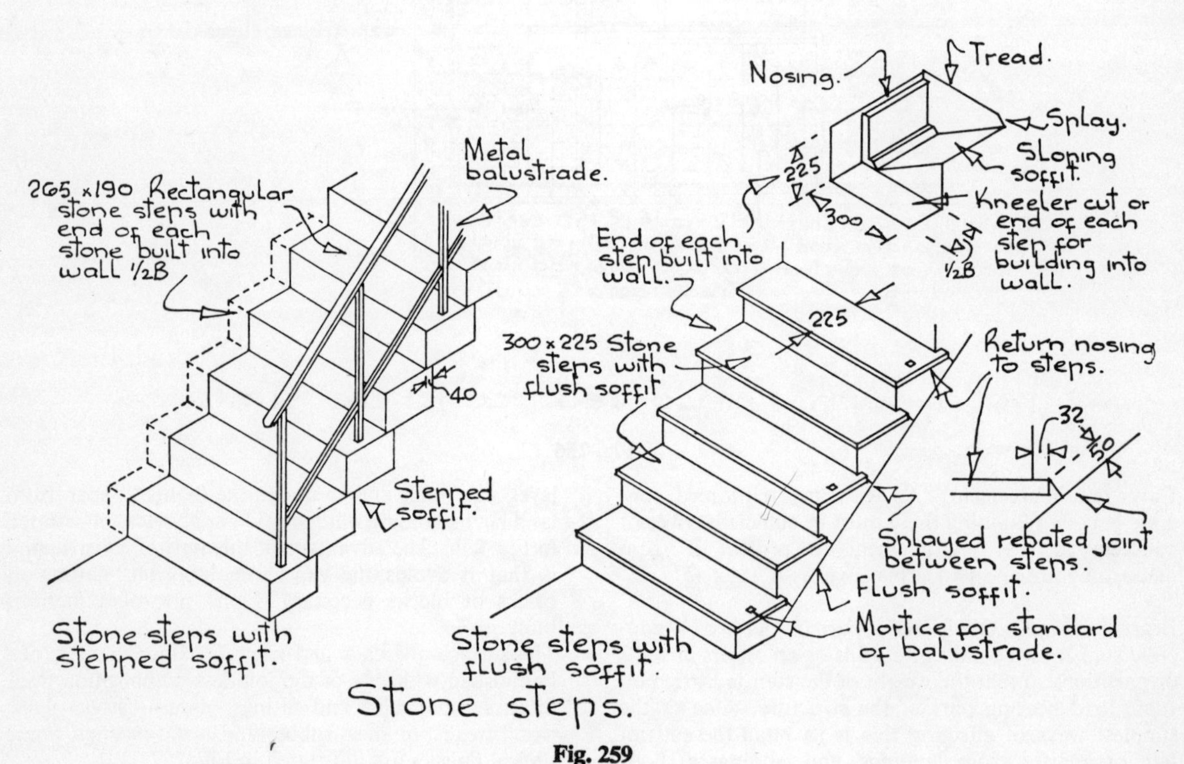


Fig. 259

### Stone stairs

These are rarely constructed today because a stone stair is as expensive to build, but not as strong, as a similar reinforced concrete stair. But before cement was manufactured the stairs of larger buildings were often constructed of natural stones. Flights were constructed with either rectangular section stones or triangular section stones one end of each stone being built into the walls of the stair well as illustrated in Fig. 259.

Landings were constructed with one or more large stone slabs whose sides were built into the enclosing walls of the stair well.

Metal balustrades were fixed by setting their balusters or standards in holes in the stone steps and landing as previously described for concrete stairs.

#### Reference Books and Publications

**British Standards:**
No. 585. Wood stairs.

CHAPTER EIGHT

# INTERNAL FINISHES AND EXTERNAL RENDERING

**SfB (42) Finishes, internal: General**
**SfB (41) Finishes, external: General**

## PLASTER

The finished surface of walls of brick, concrete, stone and concrete or clay blocks is generally so coarse textured that it is an unsuitable finish for the internal walls of most buildings. These surfaces are usually rendered smooth by the application of two or three coats of plaster. Similarly the soffit (ceiling) of concrete floors and roofs is usually rendered smooth with plaster.

It is not fashionable today to leave the joists of timber floors and roofs exposed in the rooms below and they are covered with plaster spread on timber or metal lath or with plasterboard, to provide a smooth, level ceiling and to give the timbers some protection against damage by fire.

The purpose of plaster is to provide a smooth hard level finish to walls and ceilings.

But fashions change, and it is not uncommon today for one or all of the walls of rooms in modern houses to be finished with brickwork or stonework exposed or with paint applied directly to its surface.

The finished surface of plaster should at once be flat and fine textured (smooth). It would seem logical, therefore, to spread some fine grained material such as lime, mixed with water, over the surface and trowel it smooth and level. As the lime dried out it would harden into a dry smooth surface. But this is not practical because so fine grained a material as lime, when mixed with water, cannot be spread and trowelled smooth to a thickness of much more than say 3 and the wall would absorb so much water that the lime would crack as it dried out. A thicker application of the material would sag and run down as it was being spread. Instead, some coarser-grained material is first spread on walls in one or two coats to render the surface level and when this has dried a thin coat of fine grained material is spread over it to provide a level and smooth surface.

**Plaster undercoats.** The first coat of plaster is described as the first undercoat or render coat and it consists of sand, mixed with lime or cement, or both, and water. The material is spread and struck off level with a straight edge to a thickness of about 10. If the surface on which the render coat is applied is uneven it is good practice to spread a second coat, called the second undercoat or float coat, of the same materials and mix as the render coat and finished to a thickness of about 6. Finally a coat of some fine grained material, such as lime or gypsum plaster, mixed with water, is spread and trowelled smooth to a thickness of about 3. The final thin coat, of fine material, is described as the finishing or setting coat.

To reduce the cost of plastering it has become common practice of recent years to apply only two coats of plaster, the render and set coats, (one undercoat and a finishing coat) to walls. But the finished surface of two coat plaster applied on an irregular surface is rarely as flat as that produced by three coat plaster properly applied. Because of variations in the sizes of bricks and blocks the surface of walls built with them is irregular, and it is to hide these irregularities that plaster is used. To explain the disadvantage of two coat plaster consider the surface of a wall built with very irregularly shaped bricks as illustrated in Fig. 260. If a render coat is applied and struck off level it will tend to sag as indicated by the line at "A", due to the weight of the thick, still plastic plaster at that point. This of course will occur at several places on an irregular surface, where the render coat is thickest, and in consequence the setting coat spread over it will not be finished flat. But if a thin float coat is used on top of the render coat it can be spread and finished flat as shown in Fig. 260, and then the setting coat will also be finished flat.

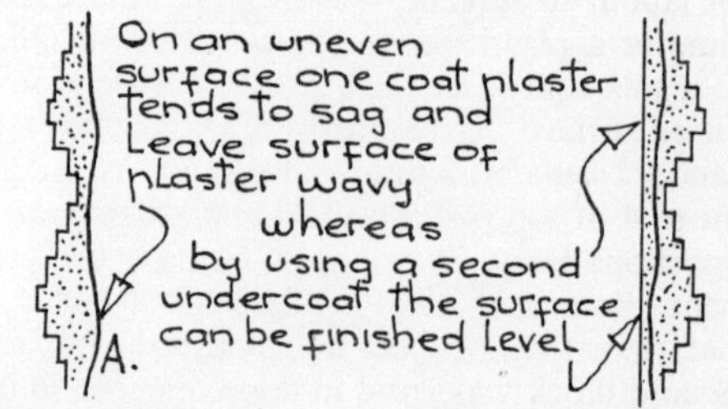


**Fig. 260**

**Materials used in plaster.** The undercoats in plaster, that is, the render and float coats, consist of some coarse grained material such as natural sand which is hard, insoluble and inert, bound with a matrix of lime or cement or gypsum.

**Lime plaster.** Before the nineteenth century Portland cement was not manufactured and the matrix then used was lime. This was mixed with sand in the proportion, 1 of lime to 3 of sand by volume and water and was termed coarse stuff by plasterers.

When a plaster of lime and sand dries out it shrinks and fine hair cracks appear on the surface. To restrain this shrinkage and to reinforce lime plaster, long animal hair was mixed in with the lime and sand; 5 kg of hair being used for every $m^3$ of coarse stuff. The resulting 'haired coarse stuff' was plastic and dried out and hardened without appreciable shrinkage. The disadvantages of lime plaster are that it is somewhat soft and easily damaged by knocks, in time it becomes dry and powdery and it is soluble in water.

Up to the end of the nineteenth century most internal plastering was executed with render and float coats of lime and sand set with a thin coat of lime and water trowelled smooth. It has been replaced by Portland cement, for undercoats, and by gypsum plaster, for both undercoats and finishing coat. Lime may be added to the Portland cement or to some types of gypsum plaster. (Note: Portland cement should not be mixed with gypsum plaster.)

**Cement plaster.** The properties of Portland cement were described in Vol. 1. It is sold as a grey powder which, when mixed with water, hardens into a solid inert mass. It is mixed with sand and water for use as the render and float coat in plaster applied to brick, concrete and clay block walls and partitions. The mix used is one part of cement to 3 or 4 parts of clean washed sand.

A mixture of cement and clean sand forms a very hard surface as it sets, but it is not plastic and requires a deal of labour to spread. It is usual therefore to add either lime or a plasticiser to the cement and sand to produce a mix that is at once plastic yet dries out to form a hard surface. Usual mixes are 1 cement, 1 lime and 6 sand; 1 cement, 2 lime and 9 sand; or a mix of 1 cement to 4 of sand with a mortar plasticiser added, the proportions being by volume. Cement and sand, or cement, lime and sand undercoats (render and float coats) are the cheapest undercoats in use today for brick and block walls and in consequence are more used than other types of undercoat.

As an undercoat of cement and sand dries out it shrinks slightly and cracks may appear in the surface. In general the more cement used the greater the shrinkage and therefore cracking. The extent of the cracking that may appear depends to some extent on the strength of the surface on which the plaster is applied and the extent to which the plaster binds to the surface. For example the surface of keyed fletton brickwork is sufficiently strong and affords sufficient key to restrain any appreciable shrinking, and therefore cracking, of plaster, but the surface of some lightweight concrete blocks is not sufficiently strong to prevent cracking of this type of plaster. A cement sand undercoat should therefore be used only on a backing of hard brick, dense aggregate concrete blocks or clay blocks; on lightweight blocks a cement, lime, sand mix should be used (Mix 1 – 1 – 6 or 1 – 2 – 9).

**Gypsum plaster**

During the last fifty years the use of gypsum plasters has increased greatly so that they have almost superseded lime as a finishing (setting) plaster and are also used as a matrix with sand for undercoats to a considerable extent. The advantage of the gypsum plasters is that they expand very slightly on setting and are not therefore likely to cause cracking as are cement and lime.

Gypsum is a chalk-like mineral mined in several parts of England. It is a crystaline combination of calcium sulphate and water ($CaSO_4\ 2H_2O$). If powdered gypsum is heated to about 170°C. it loses about three-quarters of its combined water and the resultant is described as hemi-hydrate gypsum plaster ($CaSO_4\ \frac{1}{2}\ H_2O$)-This material is better known as plaster of Paris.

If gypsum is heated at a considerably higher temperature than 170°C. it loses practically all of its combined water and the resultant is anhydrous gypsum plaster. In British Standard 1191 five classes of gypsum plaster are noted. The following is a description of the five classes.

Class A and B are based on Plaster of Paris.

**Class A. (Plaster of Paris) Hemi-hydrate Gypsum Plaster.** This material is supplied in powder form as "Plaster of Paris" or "Gauging plaster." When Plaster of Paris is mixed with water it sets so quickly (10 minutes) that it is unsuitable for use as a wall or ceiling plaster. It is used for fibrous plaster work. In fibrous plaster work the wet mix of plaster of Paris is brushed into moulds which are used for the reproduction of cornices and other decorative plaster work.

**Class B. (Board plaster) Retarded hemi-hydrate gypsum plaster.** To make plaster of Paris suitable for use as a wall or ceiling plaster the speed of its set is slowed down or retarded and this is done by adding "Keratin", an animal protein, to it in small amounts during manufacture. The amount of retarder added depends on the use of the plaster. Plaster for use in undercoats is more heavily retarded than that for use in finishing coats which require less time to spread and trowel.

This class of plaster is often known as "Board plaster" as it is the only one of the five classes of gypsum plaster which will adhere strongly to the surface of gypsum plaster board and for that reason it is used as a finishing coat and in sanded undercoats on gypsum plaster board and gypsum lath.

This plaster is also used as the matrix (binding agent) for undercoat plaster on brick, concrete and block surfaces, when it is mixed with sand in the proportions of 1 of gypsum plaster to 2½ or 3 of sand. As a finishing coat it is used neat with water. This plaster is marketed as finishing, undercoat or dual purpose plaster.

Class C. D. and E. gypsum plasters (B.S.1191) are based on anhydrous calcium sulphate.

**Class C. Anhydrous gypsum plaster (moderately burned)**
After gypsum has been burned at a high temperature to form anhydrous gypsum plaster it is ground to a fine powder and ½ to 1 % by weight of alum or zinc sulphate is added to it. These are added to the plaster to accelerate its hardening which otherwise would be so slow as to make it unsuitable for use as a wall plaster.

This class of gypsum plaster is much used as a finishing plaster on undercoats of cement, cement and lime or gypsum and sand mixes, because it is easy to work and can be trowelled to give a smooth finish. One of its useful characteristics is that it can be "brought back" (retempered). Once this plaster has been mixed with water and spread it becomes stiff but can then, or even several hours later, be made sufficiently plastic by sprinkling its surface with water, for it to be trowelled to a smooth finish. The words "brought back" describe the operation of making the plaster plastic by sprinkling it with water.

Anhydrous gypsum plaster is also used for undercoat plaster, generally with lime and sand in the proportions 2 gypsum, 1 lime, 6 sand. The plaster is marketed as finishing, undercoat or dual purpose (used either in undercoat or finishing coat).

**Class D. (Keenes or Parian) Anhydrous gypsum plaster (hard burned).** This class of gypsum plaster is made by heating gypsum beyond the point at which it becomes anhydrous moderately burned. The burned gypsum is ground to a fine powder and an accelerator is added as described for Class C. plaster. The best known plaster in this class is Keenes cement. It is easy to work, can be trowelled to give a particularly smooth hard finish and is used mostly as a finishing coat. It is more expensive than other types of gypsum plaster and is used in high-class work. As it dries out and hardens this plaster expands and should only be applied over strong undercoat plaster such as cement and sand (1 : 3) or undercoats of Keenes and sand (1 : 2). Because it is particularly hard Keenes is often used as a finishing plaster to plaster angles. Parian cement is similar to Keenes but is not manufactured today.

**Class E. Anhydrite.** Is a naturally occurring anhydrous calcium sulphate, mined in England. When crushed and mixed with an accelerator it produces a material similar to Class C. gypsum plaster.

Gypsum plaster is used as the finishing coat in most plasterwork today.

For undercoat plaster clean natural sand with lime and cement is generally used when building costs have to be kept as low as possible, and sand and gypsum plaster when cost is not a first consideration. A gypsum plaster undercoat is less likely to crack than one of lime or cement.

**Sand for plastering**
Sand for plastering should be clean and contain not more than 5 % of clay or other soluble adherent matter. In Vol. 1 an explanation was given of why sand for mortar should be clean, and this applies equally to sand for plastering. The sand should not contain any grains larger than 5.

**Lightweight aggregate for plaster.** Of recent years lightweight aggregate has been used in lieu of natural sand in plaster. The aggregates commonly used are perlite and exfoliated vermiculite. Perlite and vermiculite are minerals which expand into multicellular lightweight materials when heated. The expanded minerals, though lightweight and cellular, have sufficient mechanical strength for use as an aggregate for plaster. The description exfoliated vermiculite describes the action of the mica like mineral vermiculite, when heated. The thin layers of the mineral open up (exfoliate) when heated to form a cellular mass.

For use as an aggregate in plaster the expanded mineral is crushed until the particles are no larger than 5.

These lightweight aggregates can either be mixed with gypsum plaster on the building site or prepared mixes of the aggregate and gypsum plaster are supplied by the plaster manufacturers. Gypsum plaster Class B. (retarded hemi-hydrate) is generally used with light weight aggregates.

Two undercoats of lightweight aggregate and gypsum are generally used and spread and levelled in the same way as ordinary plaster. A finishing coat of gypsum Class B. is generally used. The advantages of these plasters are they are light in weight, being less than half the weight of plasters made with sand. Because of their cellular nature these aggregates are better thermal insulators than natural sand and the insulation of a wall or ceiling can be quite considerably improved by their use in plaster. The condensation of moisture, from warm moisture-laden air, on cold wall surfaces in rooms such as bathrooms and kitchens can be considerably reduced by the use of a lightweight aggregate plaster. This reduction of condensation is due to the insulating property of the plaster which prevents the inside face of walls being as cold as it would be if dense aggregate (sand) were used.

Lightweight aggregate plasters have good resistance to damage by fire and can be used to protect such structural parts as timber floor joists. Because of the cellular nature of the aggregate lightweight plasters are not so resistant to damage by knocks or abrasions as dense aggregate plasters. A lightweight aggregate gypsum plaster is about twice the cost, per metre of plastered surface, of a cement, lime, sand and setting coat plaster.

**Plaster finishes to timber joists and studs**
The usual method of providing a level finished surface to the ceiling (soffit) of timber floors and roofs and on timber stud partitions is to spread plaster over timber or metal lath or to fix preformed boards to the timber ceiling or wall.

**Fir lath.** Before the twentieth century the usual method of preparing timber ceilings and timber stud walls and partitions for plaster was to cover them with fir lath spaced about 7 to 10 apart to provide a key for the plaster. The usual size of lath is 25 wide by 5 to 7 thick, in lengths of 900. Lath is either split or sawn from Baltic fir (softwood). Split lath is usually described as riven lath and is prepared by splitting along the grain of the wood. Because the grain of wood is never absolutely straight neither is riven lath, so that when it is fixed the spaces left between the laths as a key for the plaster are not uniform. This may prevent the plaster being forced between the laths and it will not therefore bind firmly to them. Sawn lath on the other hand is uniformly straight and can be fixed with uniform spaces to give a good key for plaster. Fir lath must be adequately seasoned and free from fungal decay.

The fir lath is nailed across the joists or timber studs. Obviously the ends of the laths must be fixed to a joist or stud as illustrated in Fig. 261, and the butt end joints of laths staggered to minimise the possibility of cracks in the plaster occurring along the joints.

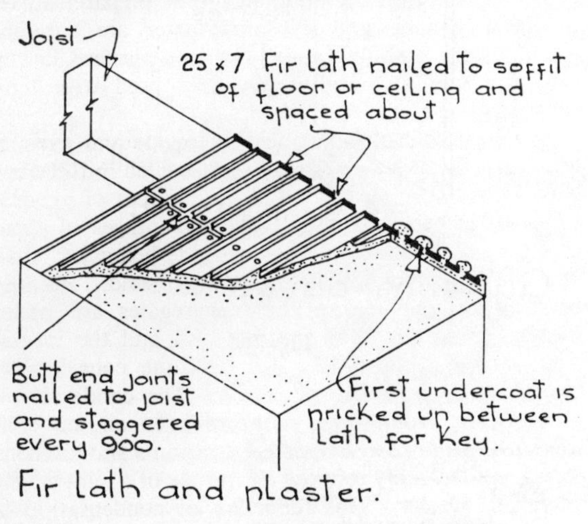


**Fig. 261**

Fir lath is covered with three coats of plaster. The first coat is spread and forced between the lath so that it binds to it. This coat is described as pricking up. A second undercoat, termed the float coat, is spread and finished level and then covered with the finish or setting coat.

Before the twentieth century the undercoats consisted of haired coarse stuff (lime : sand— 1 : 3, with hair) gauged with plaster of Paris, and the finishing coat of lime and water gauged with plaster of Paris. The purpose of the gauge (addition of a small amount) of plaster of Paris is to cause the material to harden more quickly than it would otherwise do, so that vibration due to the applications of the next coat, or vibrations of the floor above, will not cause the plaster to come away from the lath before it is hard.

Plastering on fir lath today is generally executed with a mix of cement : lime : sand (1 : 2 : 9) or gypsum plaster : lime : sand (1 : 2 : 9) for undercoats, and gypsum plaster Class C. as a finishing coat. The cost of fir lath and three coats of plaster today is about three times that of a plasterboard finish and in consequence fir lath is less used now than it was.

**Metal lath: (E.M.L.).** This lath is made by cutting thin sheets of steel so that they can be stretched into a diamond mesh of steel as illustrated in Fig. 262. This lath is described as E.M.L. (expanded metal lath). The thickness of the steel sheet which is cut and expanded for plasterwork is usually 0·8 mm, 0·6 mm, or 0·5 mm and the lath is described by its shortway mesh. A mesh of 6 or 10 shortway is generally used for plaster. To prevent expanded steel lath rusting it is either coated with paint or galvanised.

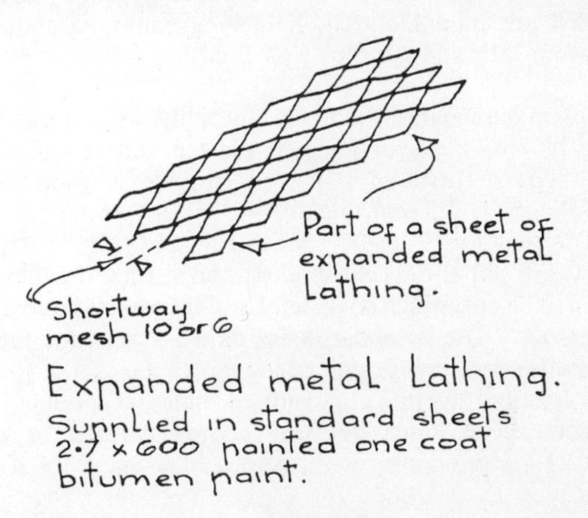


**Fig. 262**

As a background for plaster on timber joists and studs, the lath, which is supplied in sheets 2·7 x 600, is fixed by nailing with galvanised clout nails or galvanised staples at intervals of about 100 along each joist or stud. During fixing the sheet of lath should be stretched tightly across the joists. Edges of adjacent sheets of the lath should be lapped at least 25.

The undercoat plaster generally used on this type of lath is lime and sand gauged with Portland cement or gypsum. Three coat plasterwork should be used. The following are some mixes of plaster commonly used:

Undercoats: 1 : 2 : 9 : cement : lime : sand
1 : 2 : 9 : gypsum class B or C : lime : sand.

Finishing coat: Gypsum Class B. or C.

The cost of three coat plaster on metal lath is rather more than twice the cost of plasterboard finished with a skim coat.

Metal lath is principally used as a background for plaster in ceilings suspended below concrete floors and steel roofs. The lath is supported on light steel runners hung on steel hangers fixed to the floor or roof above.

**Gypsum plasterboard.** Gypsum plasterboard consists of a core of set (hard) gypsum plaster enclosed in and bonded to two sheets of heavy paper. The heavy paper protects and reinforces the gypsum plaster core which otherwise would be too brittle to handle and fix without damage. Plasterboard is made in thickness of 9·5 mm and 12·7 mm for use as a plaster finish and in boards of various sizes.

Plasterboard is very extensively used as a finish on the soffit (ceiling) of timber floors and roofs and on timber stud partitions. The advantages of this material as a finish are that it provides a cheaper finish, it can more speedily be fixed and plastered and provides better fire protection than lath and plaster. Its disadvantages are that, because it is a fairly rigid material, plaster finishes applied to it may crack due to vibration or movement in the joists to which it is fixed and it is a a poor sound insulator. Four types of gypsum plasterboard are manufactured, namely gypsum baseboard, gypsum lath and gypsum wall board.

**Gypsum baseboard:** Is made specially as a base for one or two coats of gypsum plaster and is enclosed with coarse textured paper to provide a good key. Baseboard is 9·5 mm thick, 914 wide and in lengths of 1·2 to suit joists at 400 centres, Fig. 263. Obviously if joists are not at the centre noted above then the boards have to be cut, which is wasteful and increases the cost of the work. The baseboards are nailed across the joists or studs with their bound edges across the run of joists and secured with 32 galvanised nails at about 150 centres. The joints between boards parallel to the run of joists should be broken and a space of 4 to 7 left between all adjacent edges. The joints between boards are filled with a wet mix of gypsum plaster Class B., which is also spread around the joints and into which strips of open weave hessian or jute scrim are pressed. The purpose of the scrim is to reinforce the plaster over the joints between boards. Fig. 264 illustrates the fixing of baseboard. Baseboard is finished either with one coat of neat gypsum plaster Class B., spread and trowelled to a finished thickness of 5, or it is floated with haired browning plaster and sand (1 to 1½) and finished with neat gypsum plaster Class B. The finished thickness of these two coats should be about 12.

Two-coat finish to baseboard is less likely to crack and provides greater protection against damage by fire than a one-coat finish.

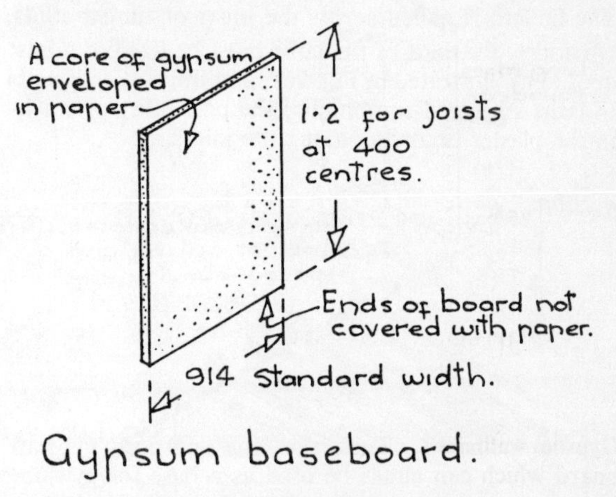


**Fig. 263**

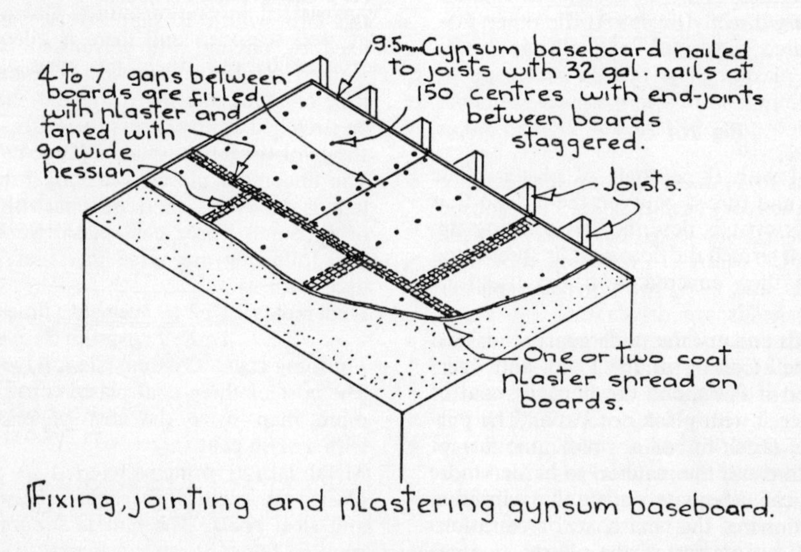


**Fig. 264**

**Gypsum lath.** Is similar to baseboard but made in smaller boards which are easier to handle and fix than baseboard. Gypsum lath is 9·5 mm thick and 406 wide and in lengths of 1·2 to suit joists at 400 centres, (Fig. 265). The lath is enclosed in heavy coarse textured paper and its long edges are rounded so that hessian scrim need not be used at the joints. The lath is nailed with its long edges across the span of joists and the end joints should be staggered. A space of about 4 is left between adjacent laths. The joints between the lath should be filled with neat gypsum Class B., and the surface finished with one or two coat plaster as described for baseboard.
Fig. 265 illustrates gypsum lath.

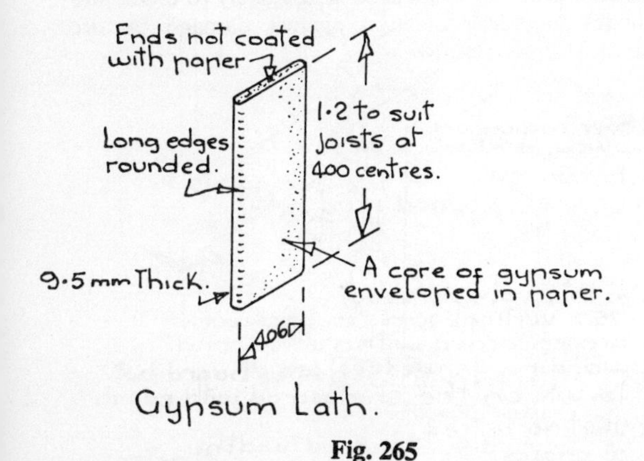


**Fig. 265**

**Gypsum wallboard.** This is a dual purpose gypsum board which can either be used as a base for gypsum plaster or as a finished surface ready for painting. The boards have a core of gypsum enclosed in heavy paper. One side of each board is coated with paper prepared for painting directly on it and the other side of the board is coated with open textured paper prepared for gypsum plaster. The boards are made in two thicknesses, 9·5 mm and 12·7 mm, in widths of 600, 900 and 1·2 and lengths of 1·8, 2·35, 2·4, 2·7 and 3·0 as illustrated in Fig. 266.

If the wall board is to be fixed as a finished surface ready for painting the ivory surface should be exposed and the boards nailed to joists or studs with 32 galvanised nails at 150 centres. A space of about 4 is left between boards and filled with gypsum plaster stopping. The nail heads are driven into the paper coating of the boards and covered with gypsum plaster stopping. When the stopping in the joints and over nail heads has dried it is rubbed down until smooth and then the surface is ready for decoration. This is the cheapest type of finish to ceilings of timber floors and to stud partitions. If the timber joists or studs are well seasoned and adequately strutted to prevent them twisting or moving this finish will not crack at the joints between boards and is particularly suitable for hanging wallpaper or ceiling paper on.

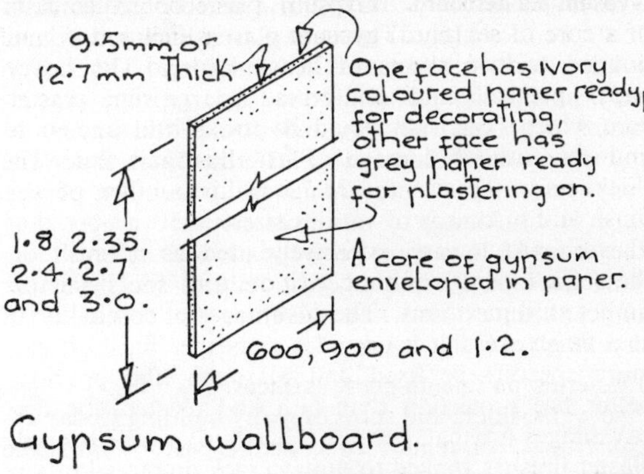

9·5 mm or 12·7 mm Thick
One face has ivory coloured paper ready for decorating, other face has grey paper ready for plastering on.
1·8, 2·35, 2·4, 2·7 and 3·0.
A core of gypsum enveloped in paper.
600, 900 and 1·2.
Gypsum wallboard.

**Fig. 266**

If the gypsum wall board is to be plastered it is nailed to joists or studs with grey surface exposed and secured with nails at 150 centres as previously described. A space of from 4 to 7 is left between boards and then either one coat or two coat plastering is executed with hessian scrim over joints as described for gypsum baseboard. If large sheets of wallboard are used, such as 3·0 x 1·2 the likelihood of cracks appearing in the finished surface is greater than with the smaller sized boards of baseboard.

**Cracking of gypsum board finishes.** The two principal causes of cracking in these finishes are (a) twisting or other movement of joists or studs to which they are fixed and (b) deflection of timber joists under load.
New timber is often not as well seasoned as it should be today. As the timbers dry out they shrink and this shrinkage may cause joists to wind (twist). As a joist winds it twists the boards fixed to it. Fig. 267 is an exaggerated illustration of this. Obviously if the timbers are well seasoned and there is adequate herringbone strutting between them this type of cracking is unlikely.

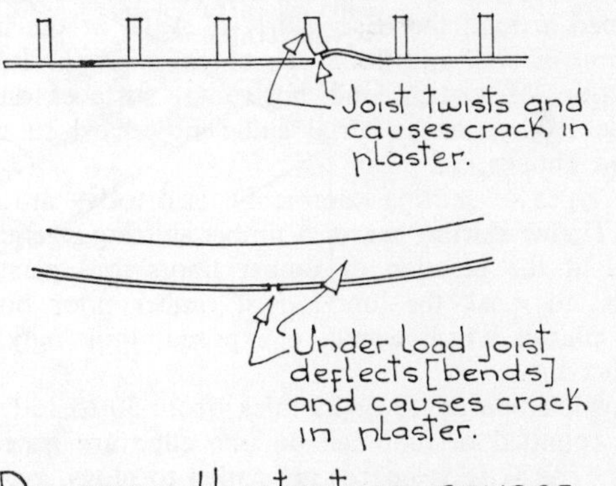


**Fig. 267**

Under the load of furniture and persons timber floor joists bend slightly. The degree to which joists bend is described as their deflection under load. Even with very small deflection under load a large rigid plasterboard may bend slightly and cracks will appear at joints between boards. Fig. 267 illustrates this. The only sure way of preventing the possibility of this type of cracking is to use joists some 50 deeper than they need be to carry expected loads with safety.
This additional depth of joist reduces the deflection of joists under load and therefore the possibility of cracking from this cause.

**Plastering on smooth dense surfaces.**
The walls, floors and roofs of many buildings today are constructed of reinforced concrete cast *in situ*. The concrete is cast inside timber or plywood formwork. When the formwork is removed and the concrete has dried, its surface is often so smooth and dense that plaster will not adhere strongly to it. One way of preparing concrete surfaces for plastering is by hacking them with a chisel and hammer or with a power operated hammer. The operation of hacking concrete is laborious and expensive.
Bonding liquids, which consist of an emulsion of polyvinyl acetate, are now made. These liquids are painted on to smooth dense surfaces, to which they adhere strongly, and when dry they provide a surface to which gypsum plaster will adhere. One coat of gypsum plaster Class B is generally used on surfaces treated with bonding liquids before plastering.
As an alternative to bonding liquids bonding plaster may be used on smooth dense surfaces. This plaster is a retarded hemi-hydrate gypsum plaster with low setting expansion, mixed with some selected wood fibres. The plaster is used neat and applied as one-coat finishing plaster. It will adhere to all but very smooth surfaces such as glazed tiles.

## SKIRTING AND ARCHITRAVES

### Skirting

The skirting is a narrow band, usually projecting, formed around the base (skirt) of walls at the intersection of wall and floor. It serves to emphasise the junction of vertical and horizontal surfaces and is made from some material sufficiently hard to withstand knocks.
The types of skirting commonly used today are:
(1) **Timber skirting board.** A timber skirting is generally used at the junction of timber floors and plastered walls, to mask the junction of timber floor boards and plaster which would, if exposed, look ugly and collect dirt.
Softwood boards, 19 or 25 thick, from 50 to 150 wide and rounded or moulded on one edge are generally used. The skirting boards are nailed to plugs, grounds or concrete fixing blocks at the base of walls after plastering is completed. Figs. 268 and 269 illustrates some typical section of skirting board and the fixing of the board.

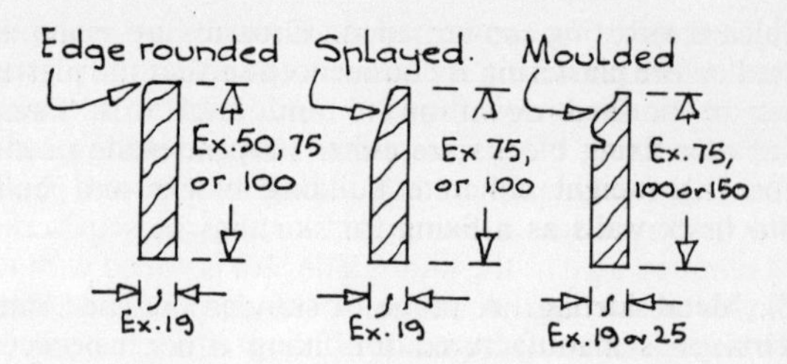


**Fig. 268**

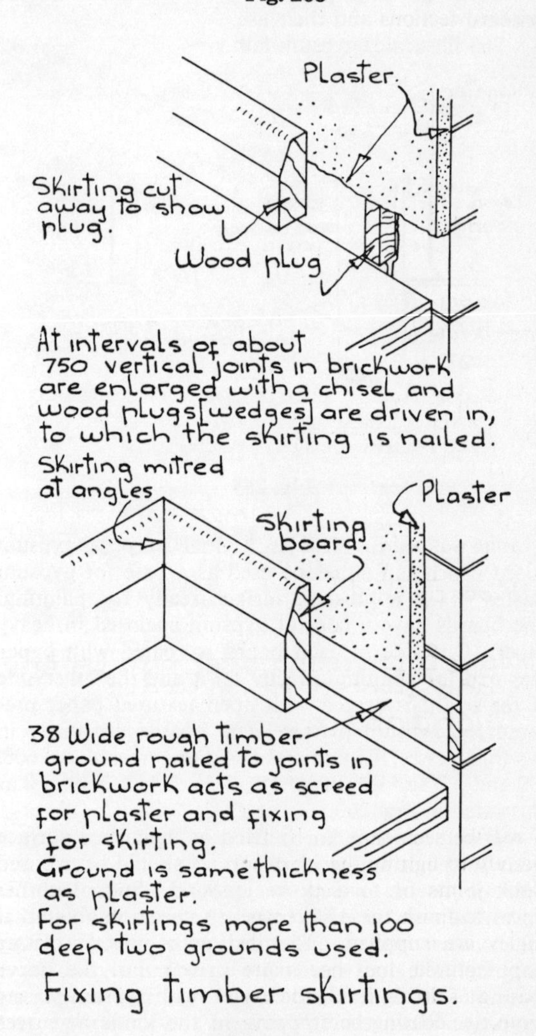


**Fig. 269**

Plugs are wedge shaped pieces of timber which are driven into brick or block joints from which the mortar has been cut out. Grounds are small section lengths of sawn softwood timber, 38 or 50 wide and as thick as the plaster on the wall. These timber grounds are nailed horizontally to the wall as a background to which the skirting can be nailed. Grounds are generally fixed before plastering is commenced so that the plaster can be finished down on to and level with them. Concrete fixing blocks are either purpose-made or cut from lightweight concrete building blocks and built into brick walls as a fixing for skirtings.

(2) **Metal skirting.** A range of standard pressed steel skirtings is manufactured for fixing either before or after plastering. The skirting is pressed from mild steel strip and is supplied painted with one coat of red oxide priming paint. Fig. 270 illustrates the standard sections and their use.

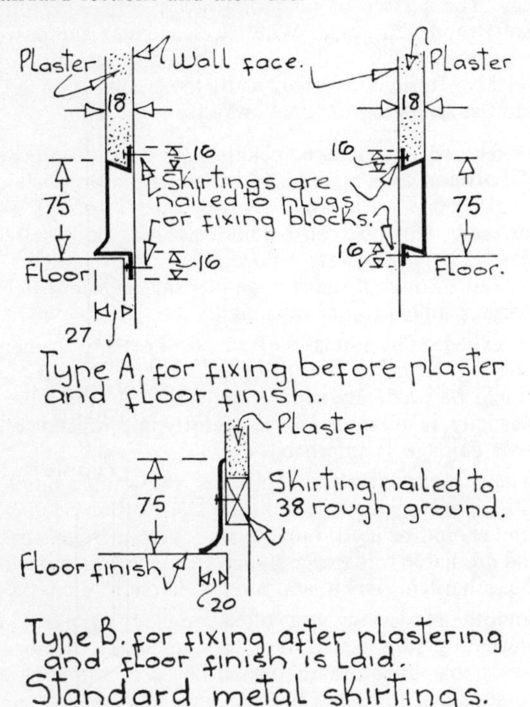


**Fig. 270**

It will be seen that the skirting is fixed by nailing it directly to lightweight blocks or to plugs in brick and block joints or to a timber ground. Special corner pieces to finish these skirtings at internal and external angles are supplied. The section of metal skirting manufactured does not make a particularly attractive finish at skirting level, the metal may rust due to the protective coating being damaged and the angle pieces are ugly, for which reasons metal skirtings are not much used.

(3) **Tile skirting.** The manufacturers of floor quarries and clay floor tiles make skirtings to match the colour and size of their products. The skirting tiles have rounded top edges and butt on to the floor finish, or they have rounded top edges and a cove base to provide an easily cleaned rounded internal angle between skirting and floor.

The skirting tiles are first thoroughly soaked in water and then bedded in sand and cement against walls and partitions. Special internal and external angle fittings are made. Fig. 271 illustrates the various types and use of these skirting tiles.

Skirting tiles make a particularly hard wearing, easily cleaned finish at the junction of floor and walls and are commonly used with quarry and clay tile finishes to solid floors.

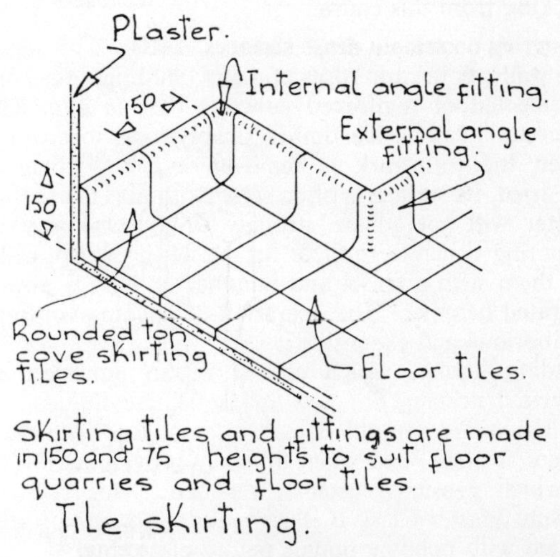


**Fig. 271**

(4) **Magnesite and Anhydrite skirtings.** When solid floors are finished with one of the jointless floor finishes such as magnesium oxychloride or anhydrite (Vol. 1. Chapter 5) it is quite usual for the material to be used as a skirting with a cove formed at the junction of floor and skirting. As with cove tile skirtings this makes a neat, easily cleaned, finish.

### Architrave

The word architrave describes a decorative moulding fixed or cut around doors and windows to emphasise and decorate the opening. An architrave can be cut or moulded on blocks of stone, concrete or clay, built around openings externally. Internal architraves usually consists of lengths of moulded timber nailed around doors and windows, An internal timber architrave serves two purposes, firstly to emphasise the opening and secondly to mask the junction of wall plaster and timber door or window frame. If an architrave is not used an ugly crack tends to open up between the back of frames and wall plaster. It is to hide this crack that the architrave is fixed.

A timber architrave is usually 19 or 25 thick and from 50 to 100 wide. It may be finished with rounded edges, or splayed into the door or decorated with some moulding. Usual practice is to fix architraves so that they diminish in section towards the door or window

Narrow architraves can be fixed by nailing them to the frame or lining of the door or window. Wide architraves are usually fixed to sawn softwood grounds nailed to the wall around the frame or lining as a background to which the architrave can be securely nailed. Architraves are mitre cut (45 deg. cut) at angles. Fig. 272 illustrates some typical sections and fixing of architraves.

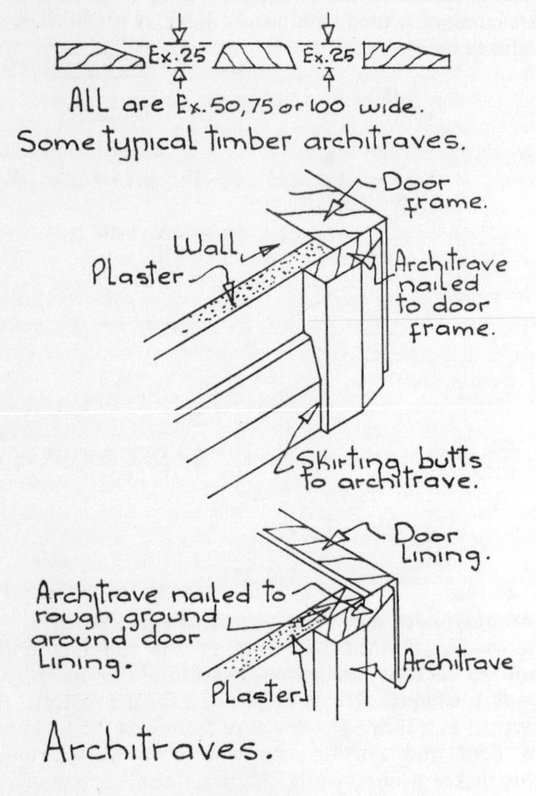


Fig. 272

## EXTERNAL RENDERING

Owing to their colour and texture, common bricks and concrete and clay blocks do not provide an attractive external finish for buildings. The external faces of walls built with these materials are commonly rendered with two or three coats of cement and lime mixed with natural aggregate and finished either smooth or textured.

In exposed positions walls may become so saturated by rain that water penetrates to their inside face. Because an external rendering generally improves the resistance of a wall to rain penetration, the walls of buildings on the coast and on high ground are often rendered externally.

The types of external rendering used are:
(1) Smooth; (2) Textured; (3) Rough cast; (4) Pebble dash.

(1) **Smooth rendering.** Is applied in two or three coats. Mixes of cement and sand, 1 : 3, or cement, lime and sand, 1 : 1 : 6, or 1 : 2 : 9, by volume, are generally used.

The surface to be rendered should afford sufficient key for the material. Keyed flettons, grooved clay blocks, and most concrete building blocks provide a surface quite satisfactory for rendering. Walls built of bricks which are not keyed should have the mortar joints raked out about 25 as they are built, to provide a key for rendering.

Sand for rendering must be clean (sharp) and free from clay. The surface of rendering mixes containing dirty (soft) sand will crack. Water which enters the cracks will saturate the wall and will be unable to dry out quickly. In time frost may cause the cracks to open up and the rendering to come away from the wall.

**Mixes used.** In sheltered positions rendering composed of Portland cement, lime and sand in the proportions 1 : 2 : 9, can successfully be used on all normal wall surfaces. This mix can readily be spread because it is plastic, it is moderately porous and allows water to dry out through it quickly, and its surface is not liable to crack unless soft sand is used.

In exposed positions a 1 : 1 : 6, Portland cement, lime, sand, or a 1 : 3, Portland cement, sand mix should be used. By using more cement in the mix its plasticity is reduced but its density and resistance to frost damage is improved.

Where renderings are applied in the winter months in exposed positions a mix of 1 : 3, Portland cement, sand should be used. Lime cement mixes harden slowly and are liable to damage by frost whereas cement sand mixes harden quickly and are less liable to damage.

Smooth renderings are often applied to assist in preventing rain penetration through walls. It seems reasonable therefore to use as dense a rendering as possible and for years builders have employed cement and sand renderings (mix 1 : 3 by volume). But a cement sand mix is not plastic if clean sand is used. Builders often therefore use soft (dirty) sand, the consequence of which is that the rendering cracks. Water enters the cracks and cannot evaporate through the dense rendering and the wall often remains more heavily saturated by rain than it would be if there were no rendering on it.

To make rendering mixes, plastic lime is used with cement and clean sand (mix 1 : 1 : 6, or 1 : 2 : 9, Portland cement, lime, sand) to combine the plasticity of lime with the density of cement.

Mortar plasticisers (see Vol. 1) are now available which, when mixed in with cement and sharp sand, make the mix sufficiently plastic for it to be readily spread and finished smooth.

Lightweight concrete blocks should not be rendered with cement—sand mixes because they are not sufficiently strong to resist the drying shrinkage due to the cement. Cement-sand renderings applied on lightweight concrete blocks will crack and come away from the blocks. One of the cement, lime, sand mixes, which do not shrink so severely, should be used. In sheltered situations a 1 : 2 : 9, cement, lime, sand mix may be used and in more exposed situations a 1 : 1 : 6, cement, lime, sand mix.

**Application of rendering.** Rendering is usually applied in two coats. The first coat is spread by trowel and struck off level to a thickness of about 12. The surface of the first coat is scratched before it dries to provide key for the next coat. The first coat should be allowed to dry out. The next coat is spread by trowel and finished smooth and level to a thickness of about 10 The surface of smooth renderings should be finished with a wood float rather than a steel trowel. A steel trowel brings water and the finer particles of cement and lime to the surface which, on drying out, shrink and cause surface cracks. A wood float (trowel) leaves the surface coarse-textured and less liable to surface cracks. Three coat rendering is used mostly in exposed positions to provide a thick protective coating to walls. The two undercoats are spread, scratched for key, and allowed to dry out to a thickness of about 10 each coat and the third or finishing coat is spread and finished smooth to a thickness of about 10

**Spatterdash.** Smooth, dense wall surfaces such as dense brick and situ-cast concrete afford a poor key and little suction for renderings. Such surfaces can be prepared for rendering by the application of a spatterdash of wet cement and sand. A wet mix of cement and clean sand (mix 1 : 2 : by volume) is thrown on to the surface and left to harden without being trowelled smooth. When dry it provides a surface suitable for the rendering which is applied in the normal way.

(2) **Textured rendering.** The colour and texture of smooth rendering appears dull and unattractive to some people and they prefer a broken or textured surface. The choice of mix used for this type of rendering depends on the exposure of the building and the nature of the surface on which the rendering is to be applied, as it does for smooth rendering.

Whether a cement-sand, or cement-lime-sand mix is chosen, textured rendering is usually applied in two coats. The first coat is spread and allowed to dry as previously described. The second coat is then spread by trowel and finished level. When this second coat is sufficiently hard, but still wet, its surface is textured by scraping or rubbing it with wood combs, brushes, sacking, wire mesh or old saw blades. A variety of textures can be obtained by varying the way in which the surface scraping or rubbing is carried out.

An advantage of textured rendering is that the surface scraping removes any scum of water, cement and lime that may have been brought to the surface by trowelling and which might otherwise have caused surface cracking.

Heavily textured surfaces are not generally satisfactory in industrial atmospheres as soot and rain causes irregular unsightly surface stains.

(3) **Rough cast.** This is a form of textured rendering, the texture being achieved by throwing the finishing material on to the surface either by hand or by machine. Two undercoats are spread and levelled by trowel, and allowed to dry. The finishing coat is then thrown on. The undercoat mix is either cement, lime, sand: 1 : 1 : 6, or cement and sand : 1 : 3. The finish is a mixture of cement-lime-sand and gravel, or coloured cement-sand and gravel mixes, the former applied on cement-lime-sand undercoats and the latter on cement-sand undercoats. Roughcast finishes applied by throwing on by hand have a rough somewhat irregular surface. Finishes applied by machine have a regular rough surface.

Coloured cement, sand and gravel finishes are described as Tyrolean or Tyrolean cement finishes.

(4) **Pebble dash.** This is another form of textured finish achieved by throwing clean pebbles or broken stone on to the still wet surface of the second undercoat of the rendering.

A first undercoat of cement, sand : 1 : 3 is spread, levelled and allowed to dry out. The second undercoat of the same mix is spread and levelled and while it is still wet the selected clean small gravel pebble or broken stone is thrown on until the whole undercoat surface is covered. The finished effect is a coarse surface of exposed particles of gravel or broken stone.

This type of finish is popular because the surface of stone particles masks any cracks that might otherwise be visible and the hard smooth surface of the exposed pebbles does not become soot-encrusted in any but the most soot-laden atmospheres.

**Reference Books and Publications**

**British Standards:**
No. 12. Portland cement.
No. 584. Wood trim (skirtings and architraves).
No. 890. Building limes.
No. 1191. Gypsum building plasters.
No. 1198. Sands for plastering with gypsum.
No. 1199. Sands for plastering with lime and cement.
No. 1230. Gypsum plasterboard.
No. 1246. Metal skirtings, picture rails and beads.
No. 1286. Clay tiles for flooring.
No. 1317. Wood laths for plastering.
No. 1369. Metal lathing for plastering.
**British Standard Code of Practice:**
CP. 211. Internal plastering.
**Building Research Station Digests:**
No. 4. Pattern staining in buildings. (First Series).
No. 26. Blowing, popping or pitting of internal plaster. (First Series).
No. 49. Choosing specifications for plastering.

# INDEX


A

Aberdeen granite, 1
Ancaster stone, 3
Angle block, 120
Anhydrite, 130
Anhydrous plaster, 130
Anti-capillary groove, 36
Apron, 92, 94
Arches, 16
 flat, 15
Architrave, 135
Artificial stone, 22
Ashlar masonry, 10

B

Back, boiler, 102
 gutter, 94
 putty, 50
Baluster, 122
Balustrade, 121, 126
Barefaced tenon, 80
Basebed, 2
Bath stone, 2
Bead, butt, 67
 flush, 68
Bed of stone, 17
Bedding stone, 4
Beer stone, 3
Bevel, raised, 66
Blast furnace slag, 107
Blaxter stone, 3
Block, clay, 108
 concrete, 106
 flue, 104
Board plaster, 129
Bond stone, 4
Bonding liquid, 134
Bolection mould, 66
Bottom rail, 35, 60
Box ground stone, 2
Brace, 79
Bracket, 123
Brass butt, 82
Breast, 87
Breeze block, 108
Brick sill, 54
Building in lug, 31, 71
Bullnose step, 123
Butt, brass, 82
 cast iron, 82
 double pressed, 81
 rising, 82
 skew, 82
 steel, 81

C

Carriage, 123
Cased frame, 27, 46
Casement, 24, 28, 35, 38, 40
Casement door, 69
Cast, concrete sill, 55
 iron butt, 82
 stone, 22
 sill, 56
Catch, 30
Cavity stone wall, 8, 13
Cement, joggle, 18
 plaster, 129
Chamfered joint, 20
Channelled joint, 20
Chimney, 85
 breast, 87
 pot, 90
 stack, 89
Clay block, 108
Clayware sill, 57
Clinker, 107
Clipsham stone, 3
Closed string, 119
Combed joint, 40
Concrete, block, 106
 stair, 124
 tile sill, 59
Condensation, flue, 103
Convector fire, 101
Coping, 17
Corbel, 96
Corsham Down stone, 2
Cornice, 17
Cornish granite, 2
Cottage window, 28
Cradling piece, 96
Cramp, 112
 metal, 20
 slate, 20
Crosland Hill stone, 3
Crossetted voussoir, 17
Crow Hall stone, 3
Cut string, 120
Cylinder latch, 83

D

Darley Dale stone, 3
Dead light, 26, 40
Devon granite, 2
Diatomaceous earth, 108
Diminishing stile, 68
Doddington stone, 3
Dogleg stair, 117
Door, flush, 60
 frame, 70
 glazed, 68
 lining, 70, 75
 matchboarded, 60
 panelled, 60
 stop, 75
Double glazing, 51
Double hung sash, 27
Double pressed butt, 81
Double tenon, 61
Doulting stone, 3
Dowel, wood, 62, 70
 slate, 18,
Dowelled joint, 63
D.P.C., 12, 90
Drip mould, 39
Drop newel, 121
Dry filling, 95, 100

E

Electro-galvanising, 32
E.M.L., 131
Enclosed balustrade, 121
Enclosed fire, 102
Expanded clay, 107
Extrados, 16

F

Facing, ashlar, 10
Faience sill, 58
Fastener, window, 30
Felspar, 1
Fender wall, 95
Field (panels), 66
Finishing coat, 128
Fir lath, 131
Fire check door, 78
Fire, convector, 101
 enclosed, 102
 slow burning, 101
 sunk hearth, 102
 surround, 102
Fire grate, 101
Fireback, 86, 98
Fireplace, 85, 98
Flashing, 92
Flat arch, 15
Flaunching, 91
Flight, 116
Flint wall, 8
Float coat, 128
Flue, 87

Flue, block, 104
liner, 103
Flush door, 60, 76
Fly ash, 107
Forest of Dean stone, 3
Foundations, 12, 109
Frame, door, 78
window, 24, 35
Framed door, 80
French casement, 69
Fret, 101
Furnace clinker, 107

G

Galvanising, 32
Gathering, in, 88
over, 88
Glass, 49
clear sheet, 49
polished, plate, 49
translucent, 49
Glazed door, 68
Glazing, 49
bar, 28, 69
bead, 50
clip, 50
Going, 116
Granite, 1
Grate, 101
Groove, 55
anti-capillary, 36
Grounds, 134
Gunstock stile, 68
Gypsum, 129
Gypsum lath, 133

H

Half depth joint, 96
Half turn stair, 117
Handrail, 122
Hardware, 80
Haunch, 71
Haunched tenon, 61
Head of frame, 35
Headroom, stair, 116
Hearth, 85
dimensions, 95
Hemi-hydrate plaster, 129
Hessian scrim, 132
Hinge, 80
hook and band, 82
tee, 82
Hollow clay block, 108
Hook and band, 82
Hopton Wood stone, 3
Horn, 37, 46, 71
Hot dip galvanising, 32
Housing, 119

I

Igneous stone, 1

J

Jamb, fireplace, 87
masonry, 13
Joggle, cement, 18
stone, 12, 15
Joinery, 40
Joint, casement, 36
chamfered, 20
channelled, 20
rebated, 20
Vee, 20
Joint, combed, 40
dowelled, 63
mortice and tenon, 61, 70
tongued, 77
half depth, 96

K

Keenes, 130
Kentish Rag stone, 3, 7
Knapping, 8
Knee, 98, 100

L

Lacing course, 8
Landing, 116, 121
Latch, 82
Lath, 131
Ledges, 79
Light, dead, 26
Lightweight, aggregate, 108, 130
Lightweight, concrete, 108
Lime plaster, 129
Limestone, 1
Liner, 103
Lining, 46, 70, 75
Lintel, fireplace, 86
partition, 106
stone, 14
Lipping, 77
Loadbearing partition, 106
Lock, mortice, 83
mortice dead, 83
rim, 83
Lock block, 77
Lock stile, 19
Lug, 31, 71

M

Masonry, ashlar, 10
rubble, 4
Matchboard, 79
Matchboarded door, 60, 79
Meeting rail, 46
Metal casement, 28
Metal door frame, 72
lath, 131
pin, 41
skirting, 135
window lug, 31
Metamorphic stone, 1
Mica, 1
Middle rail, 60

Mortar, block, 111
masonry, 12
Mortice and tenon, 36, 61, 62, 70
Mortice dead lock, 83
Mortice lock, 83
Mullion, 25, 36, 39
Muntin, 65

N

Natural clay sill, 57
Newel drop, 121
Newel post, 121
Noggin piece, 113
Nominal sizes, 35
Non-load-bearing partition, 106
Nosing, 116, 119

O

Open balustrade, 121
Open string, 120
Open well, 117
Openings in masonry, 13

P

Panel, 63
Panel, bead butt, 67
bead flush, 68
Panelled door, 60
Parapet, 17
Parian, 130
Parting, bead, 46
slip, 46
Partition, support, 113
Pebble dash, 136
Peg stay, 30
Perlite, 130
Pin top edge, 113
Pinned mortice and tenon, 62
Pitch, 116
Pitched face, 20
Pivoted sash, 26, 41-45
Plaster, 128
Plasterboard, 132
Plasterboard infilling, 79
Plaster of Paris, 129
Plugs, 134
Plywood, 63
facing, 77, 79
Pockets, 111
Polygonal walling, 7
Portland stone, 2
Pots, chimney, 90
Pulley, 27, 46
Pulley stile, 46
Pumice, 107
Putty, 49

Q

Quarry sill, 53
Quarter turn, 117
Quartz, 1
Quoin, 13
Quoin stone, 4

**R**

Rail, 35
Raised panel, 66
Random rubble, 4
Rebate, 36
Rebated jamb, 46
    joint, 20
Reconstructed stone, 22
Render, 128
Rendering, 136
Retarded hemi-hydrate, 129
Reticulated, 20
Rim lock, 83
Rise, 116
Riser, 116
Rising butt, 82
Roach, 2
Rock face, 20
Rough cast, 136
Rounded step, 123
Rubble cavity wall, 8
Rubble wall, 4
Rusticated, 20
Rustproofing, 31

**S**

Saddle joint, 18
Saddle piece, 92
Sand, for plaster, 130
Sandstone, 3
Sash, balance, 48
    cord, 47
    pivoted, 41
    sliding, 27, 46
Scotia, 120
Scrim, 132
Seasoning, stone, 4
Secret joggle, 15
Sedimentary stone, 1
Set, 128
Shaped step, 123
Sheradizing, 32
Shoulder, 36, 71
Sill, brick, 54
    cast concrete, 55
    cast stone, 56
    clayware, 57
    concrete tile, 59
    door, 73
    faience, 58
    natural clay, 57
    slate, 54
    stone, 55
    stoneware, 57
    terra cotta, 57
    tile, 52
    window board, 59
Skeleton cove, 76
Skew butt, 82
Skirting, 134
Slag, 107
Slate dowel, 18
Slate sill, 54
Sliding sash, 27, 46
Slot mortice and tenon, 70
Slow burning fire, 101
Smooth vendering, 136
Snapping, 8
Snecked rubble, 7
Soaker, 92
Soffit, flush, 126
Soffit, stepped, 126
Soffit lining, 46
Solid core, 78
Spatterdash, 137
Sprig, 50
Squared rubble, 6
Stack, chimney, 89
    height, 90
Staff bead, 46
Staircase, 117
Standard, door, 65, 76
    frame, 72
    metal casement, 28
    wood casement, 38
    section, 28
Stay, peg, 30
Steel butt, 82
    subframe, 33
    window sill, 34
    windowboard, 34
Stepped flashing, 92
Stepped foundation, 110
Stile, 35, 60
Stone, artificial, 22
    cast, 22
    reconstructed, 22
Stone sill, 55
    stair, 127
Stonewave sill, 57
Stooled end, 55
Stop, door, 75
Storey frame, 76, 112
Stove, 102
Straight flight, 117
Stratified stone, 2
String, 118
Stuck moulding, 64
Stud partition, 117
Subframe, 32
Sunk hearth fire, 102
Surround, fire, 102

**T**

Tee hinge, 80, 82
Tenon, 36
    barefaced, 80
    double, 61
    tusk, 96
Terminal, 90
Terra cotta sill, 57
Textured rendering, 136
Threshold, 73
Throat unit, 100
Tile skirting, 135
    sill, 52
Tiled surround, 100
Timber, sizes, 35
Timber stud, 113
Timber subframe, 33
Tongue, 63, 120
Tongued joint, 77
Tooled finisher, 20
Top rail, 35, 60
Transom, 25, 35
Tread, 116
Trimmed, joist, 96
Trimmer joist, 96
Trimming joist, 96
Tusk tenon, 96
Tyrolean cement, 137

**U**

Undercoat, 128
Universal section, 45

**V**

Vee-joint, 20
Ventlight, 25
Vermiculated, 20
Vermiculite, 130
Vertically sliding sash, 27
Voussoir, Stone, 16

**W**

Wall, rubble, 4
    thickness, 6
Wallboard, 133
Water bar, 74
Weatherboard, 36, 74
Weathered putty, 50
Weathering, cornice, 19
    stack, 92
Wedges, 36, 119
Whitbed, 2
Winders, 124
Windowboard, wood, 59
    steel, 34
Window, minimum size, 24
Window size, metal, 29
    wood, 38
Wood casement, 35
Wood skirting, 134

**Y**

York stone, 3
Yorkshire light, 28

**Z**

Z-range windows, 29
Zinc spraying, 32